SpringerBriefs in Mathematics

SpringerBriefs in Mathematics showcases expositions in all areas of mathematics and applied mathematics. Manuscripts presenting new results or a single new result in a classical field, new field, or an emerging topic, applications, or bridges between new results and already published works, are encouraged. The series is intended for mathematicians and applied mathematicians.

More information about this series at http://www.springer.com/series/10030

Stephanie Alexander · Vitali Kapovitch
Anton Petrunin

An Invitation to Alexandrov Geometry

CAT(0) Spaces

Stephanie Alexander
Department of Mathematics
University of Illinois
Urbana, IL, USA

Anton Petrunin
Department of Mathematics
Pennsylvania State University
University Park, PA, USA

Vitali Kapovitch
Department of Mathematics
University of Toronto
Toronto, ON, Canada

ISSN 2191-8198 ISSN 2191-8201 (electronic)
SpringerBriefs in Mathematics
ISBN 978-3-030-05311-6 ISBN 978-3-030-05312-3 (eBook)
https://doi.org/10.1007/978-3-030-05312-3

Library of Congress Control Number: 2018963986

Mathematics Subject Classification (2010): 53C23, 53C20, 53C45, 53C70, 97G10, 51F99, 51K10

This Springer imprint is published by the registered company Springer Nature Switzerland AG.
The registered company address is: Gewerbestrasse 11, 6330 Cham, Switzerland

Preface

This short monograph arose as an offshoot of the book on Alexandrov geometry we have been writing for a number of years. The notes were shaped in a number of lectures given by the third author to undergraduate students at different occasions at the MASS program at Penn State University and the Summer School "Algebra and Geometry" in Yaroslavl.

The idea is to demonstrate the beauty and power of Alexandrov geometry by reaching interesting applications and theorems with a minimum of preparation.

In Chapter 1, we discuss necessary preliminaries.

In Chapter 2, we discuss the Reshetnyak gluing theorem and apply it to a problem in billiards which was solved by Dmitri Burago, Serge Ferleger, and Alexey Kononenko.

In Chapter 3, we discuss the Hadamard–Cartan globalization theorem, and apply it to the construction of exotic aspherical manifolds introduced by Michael Davis.

In Chapter 4, we discuss examples of Alexandrov spaces with curvature bounded above. This chapter is based largely on work of Samuel Shefel on non-smooth saddle surfaces.

Here is a list of some sources providing a good introduction to Alexandrov spaces with curvature bounded above, which we recommend for further information; we will not assume familiarity with any of these sources.

- The book by Martin Bridson and André Haefliger [18];
- Lecture notes of Werner Ballmann [13];
- Chapter 9 in the book [20] by Dmitri Burago, Yuri Burago, and Sergei Ivanov.

Early history of Alexandov geometry

The idea that the essence of curvature lies in a condition on quadruples of points apparently originated with Abraham Wald. It is found in his publication on "coordinate-free differential geometry" [66] written under the supervision of Karl

Menger; the story of this discovery can be found in [43]. In 1941, similar definitions were rediscovered independently by Alexandr Danilovich Alexandrov; see [7]. In Alexandrov's work the first fruitful applications of this approach were given. Mainly:

- Alexandrov's embedding theorem—*metrics of nonnegative curvature on the sphere, and only they, are isometric to closed convex surfaces in Euclidean 3-space.*
- Gluing theorem, which tells when the sphere obtained by gluing of two disks along their boundaries, has nonnegative curvature in the sense of Alexandrov.

These two results together gave a very intuitive geometric tool for studying embeddings and bending of surfaces in Euclidean space, and changed this subject dramatically. They formed the foundation of the branch of geometry now called *Alexandrov geometry.*

The study of spaces with curvature bounded above started later. The first paper on the subject was written by Alexandrov; it appeared in 1951; see [8]. It was based on the work of Herbert Busemann, who studied spaces satisfying a weaker condition [24].

Yurii Grigorievich Reshetnyak proved fundamental results about general spaces with curvature bounded above, the most important of which is his gluing theorem. An equally important theorem is the Hadamard–Cartan theorem (globalization theorem). These theorems and their history are discussed in chapters 2 and 3.

Surfaces with upper curvature bounds were studied extensively in the 50s and 60s, and are by now well understood; see the survey [57] and the references therein.

Manifesto of Alexandrov geometry

Alexandrov geometry can use "back to Euclid" as a slogan. Alexandrov spaces are defined via axioms similar to those given by Euclid, but certain equalities are changed to inequalities. Depending on the sign of the inequalities, we get Alexandrov spaces with *curvature bounded above* or *curvature bounded below.* The definitions of the two classes of spaces are similar, but their properties and known applications are quite different.

Consider the space $\mathcal{M}_4$ of all isometry classes of 4-point metric spaces. Each element in $\mathcal{M}_4$ can be described by 6 numbers—the distances between all 6 pairs of its points, say $\ell_{i,j}$ for $1 \leqslant i < j \leqslant 4$ modulo permutations of the index set (1, 2, 3, 4). These 6 numbers are subject to 12 triangle inequalities; that is,

$$\ell_{i,j} + \ell_{j,k} \geqslant \ell_{i,k}$$

holds for all i, j and k, where we assume that $\ell_{j,i} = \ell_{i,j}$ and $\ell_{i,i} = 0$.

Consider the subset $\mathcal{E}_4 \subset \mathcal{M}_4$ of all isometry classes of 4-point metric spaces that admit isometric embeddings into Euclidean space. The complement $\mathcal{M}_4 \backslash \mathcal{E}_4$ has two connected components.

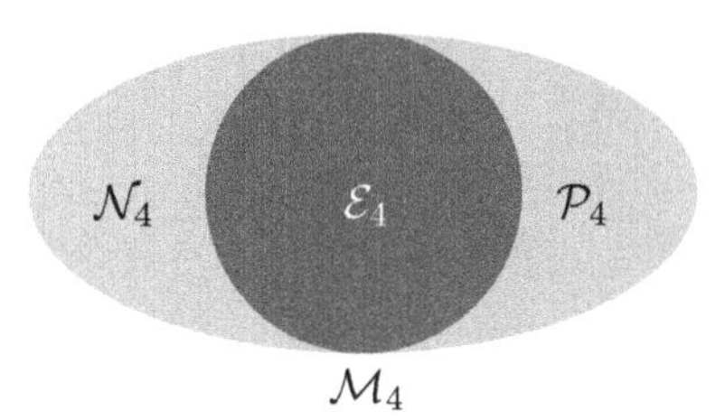

0.0.1. Exercise. Prove the latter statement.

One of the components will be denoted by $\mathcal{P}_4$ and the other by $\mathcal{N}_4$. Here $\mathcal{P}$ and $\mathcal{N}$ stand for *positive* and *negative curvature* because spheres have no quadruples of type $\mathcal{N}_4$ and hyperbolic space has no quadruples of type $\mathcal{P}_4$.

A metric space, with length metric, that has no quadruples of points of type $\mathcal{P}_4$ or $\mathcal{N}_4$ respectively is called an Alexandrov space with nonpositive or nonnegative curvature, respectively.

Here is an exercise, solving which would force the reader to rebuild a considerable part of Alexandrov geometry.

0.0.2. Advanced exercise. Assume $\mathcal{X}$ is a complete metric space with length metric, containing only quadruples of type $\mathcal{E}_4$. Show that $\mathcal{X}$ is isometric to a convex set in a Hilbert space.

In fact, it might be helpful to spend some time thinking about this exercise before proceeding.

In the definition above, instead of Euclidean space one can take hyperbolic space of curvature -1. In this case, one obtains the definition of spaces with curvature bounded above or below by -1.

To define spaces with curvature bounded above or below by 1, one has to take the unit 3-sphere and specify that only the quadruples of points such that each of the four triangles has perimeter less than $2 \cdot \pi$ are checked. The latter condition could be considered as a part of the *spherical triangle inequality*.

Urbana, USA — Stephanie Alexander
Toronto, Canada — Vitali Kapovitch
University Park, USA — Anton Petrunin

Acknowledgements

We want to thank David Berg, Richard Bishop, Yurii Burago, Maxime Fortier Bourque, Sergei Ivanov, Michael Kapovich, Bernd Kirchheim, Bruce Kleiner, Nikolai Kosovsky, Greg Kuperberg, Nina Lebedeva, John Lott, Alexander Lytchak, Dmitri Panov, Stephan Stadler, Wilderich Tuschmann, and Sergio Zamora Barrera for a number of discussions and suggestions.

We thank the mathematical institutions where we worked on this material, including BIRS, MFO, Henri Poincaré Institute, University of Colone, Max Planck Institute for Mathematics.

The first author was partially supported by the Simons Foundation grant #209053. The second author was partially supported by a Discovery grant from NSERC and by the Simons Foundation grant #390117. The third author was partially supported by the NSF grant DMS 1309340 and the Simons Foundation #584781.

Contents

Chapter 1
Preliminaries

In this chapter we fix some conventions and recall the main definitions. The chapter may be used as a quick reference when reading the book.

To learn background in metric geometry, the reader may consult the book of Dmitri Burago, Yuri Burago, and Sergei Ivanov [20].

1.1 Metric spaces

The distance between two points x and y in a metric space $\mathcal{X}$ will be denoted by $|x - y|$ or $|x - y|_{\mathcal{X}}$. The latter notation is used if we need to emphasize that the distance is taken in the space $\mathcal{X}$.

The function

$$\operatorname{dist}_x \colon y \mapsto |x - y|$$

is called the *distance function* from x.

- The *diameter* of a metric space $\mathcal{X}$ is defined as

$$\operatorname{diam} \mathcal{X} = \sup \{\, |x - y|_{\mathcal{X}} : x, y \in \mathcal{X} \,\}.$$

- Given $R \in [0, \infty]$ and $x \in \mathcal{X}$, the sets

$$\begin{aligned} \mathrm{B}(x, R) &= \{y \in \mathcal{X} \mid |x - y| < R\}, \\ \overline{\mathrm{B}}[x, R] &= \{y \in \mathcal{X} \mid |x - y| \leqslant R\} \end{aligned}$$

are called, respectively, the *open* and the *closed balls* of radius R with center x.

S. Alexander et al., *An Invitation to Alexandrov Geometry*,
SpringerBriefs in Mathematics, https://doi.org/10.1007/978-3-030-05312-3_1

Again, if we need to emphasize that these balls are taken in the metric space $\mathcal{X}$, we write

$$\mathrm{B}(x,R)_{\mathcal{X}} \quad \text{and} \quad \overline{\mathrm{B}}[x,R]_{\mathcal{X}}.$$

A metric space $\mathcal{X}$ is called *proper* if all closed bounded sets in $\mathcal{X}$ are compact. This condition is equivalent to each of the following statements:

1. For some (and therefore any) point $p \in \mathcal{X}$ and any $R < \infty$, the closed ball $\overline{\mathrm{B}}[p, R] \subset \mathcal{X}$ is compact.
2. The function $\mathrm{dist}_p \colon \mathcal{X} \to \mathbb{R}$ is proper for some (and therefore any) point $p \in \mathcal{X}$; that is, for any compact set $K \subset \mathbb{R}$, its inverse image $\{x \in \mathcal{X} : |p - x|_{\mathcal{X}} \in K\}$ is compact.

1.1.1. Exercise. *Let K be a compact metric space and*

$$f \colon K \to K$$

be a distance nondecreasing map. Prove that f is an isometry.

1.2 Constructions

Product space. Given two metric spaces $\mathcal{U}$ and $\mathcal{V}$, the *product space* $\mathcal{U} \times \mathcal{V}$ is defined as the set of all pairs (u, v) where $u \in \mathcal{U}$ and $v \in \mathcal{V}$ with the metric defined by formula

$$|(u^1, v^1) - (u^2, v^2)|_{\mathcal{U}\times\mathcal{V}} = \sqrt{|u^1 - u^2|^2_{\mathcal{U}} + |v^1 - v^2|^2_{\mathcal{V}}}.$$

Cone. The *cone* $\mathcal{V} = \mathrm{Cone}\,\mathcal{U}$ over a metric space $\mathcal{U}$ is defined as the metric space whose underlying set consists of equivalence classes in $[0, \infty) \times \mathcal{U}$ with the equivalence relation "$\sim$" given by $(0, p) \sim (0, q)$ for any points $p, q \in \mathcal{U}$, and whose metric is given by the cosine rule

$$|(p, s) - (q, t)|_{\mathcal{V}} = \sqrt{s^2 + t^2 - 2 \cdot s \cdot t \cdot \cos\alpha},$$

where $\alpha = \min\{\pi, |p - q|_{\mathcal{U}}\}$.

The point in the cone $\mathcal{V}$ formed by the equivalence class of $0 \times \mathcal{U}$ is called the *tip of the cone* and is denoted by 0 or $0_{\mathcal{V}}$. The distance $|0 - v|_{\mathcal{V}}$ is called the norm of v and is denoted by $|v|$ or $|v|_{\mathcal{V}}$.

Suspension. The *suspension* $\mathcal{V} = \mathrm{Susp}\,\mathcal{U}$ over a metric space $\mathcal{U}$ is defined as the metric space whose underlying set consists of equivalence classes in $[0, \pi] \times \mathcal{U}$ with the equivalence relation "$\sim$" given by $(0, p) \sim (0, q)$ and $(\pi, p) \sim (\pi, q)$ for any points $p, q \in \mathcal{U}$, and whose metric is given by the spherical cosine rule

$$\cos|(p, s) - (q, t)|_{\mathrm{Susp}\,\mathcal{U}} = \cos s \cdot \cos t - \sin s \cdot \sin t \cdot \cos\alpha,$$

where $\alpha = \min\{\pi, |p - q|_{\mathcal{U}}\}$.

The points in $\mathcal{V}$ formed by the equivalence classes of $0 \times \mathcal{U}$ and $\pi \times \mathcal{U}$ are called the *north* and the *south poles* of the suspension.

1.2.1. Exercise. *Let $\mathcal{U}$ be a metric space. Show that the spaces*

$$\mathbb{R} \times \operatorname{Cone} \mathcal{U} \quad \textit{and} \quad \operatorname{Cone}[\operatorname{Susp} \mathcal{U}]$$

are isometric.

1.3 Geodesics, triangles, and hinges

Geodesic. Let $\mathcal{X}$ be a metric space and $\mathbb{I}$ be a real interval. A globally isometric map $\gamma\colon \mathbb{I} \to \mathcal{X}$ is called a *geodesic*[1]; in other words, $\gamma\colon \mathbb{I} \to \mathcal{X}$ is a geodesic if

$$|\gamma(s) - \gamma(t)|_{\mathcal{X}} = |s - t|$$

for any pair $s, t \in \mathbb{I}$.

We say that $\gamma\colon \mathbb{I} \to \mathcal{X}$ is a geodesic from point p to point q if $\mathbb{I} = [a, b]$ and $p = \gamma(a), q = \gamma(b)$. In this case the image of γ is denoted by $[pq]$ and with an abuse of notations we also call it a *geodesic*. Given a geodesic $[pq]$, we can parametrize it by distance to p; this parametrization will be denoted by $\operatorname{geod}_{[pq]}(t)$.

We may write $[pq]_{\mathcal{X}}$ to emphasize that the geodesic $[pq]$ is in the space $\mathcal{X}$. We also use the following shortcut notation:

$$]pq[= [pq]\backslash\{p, q\}, \qquad]pq] = [pq]\backslash\{p\}, \qquad [pq[= [pq]\backslash\{q\}.$$

In general, a geodesic between p and q need not exist and if it exists, it need not be unique. However, once we write $[pq]$ we mean that we have made a choice of geodesic.

A metric space is called *geodesic* if any pair of its points can be joined by a geodesic.

A *geodesic path* is a geodesic with constant-speed parametrization by $[0, 1]$. Given a geodesic $[pq]$, we denote by $\operatorname{path}_{[pq]}$ the corresponding geodesic path; that is,

$$\operatorname{path}_{[pq]}(t) \stackrel{\text{def}}{=\!=} \operatorname{geod}_{[pq]}(t \cdot |p - q|).$$

A curve $\gamma\colon \mathbb{I} \to \mathcal{X}$ is called a *local geodesic* if for any $t \in \mathbb{I}$, there is a neighborhood U of t in $\mathbb{I}$ such that the restriction $\gamma|_U$ is a geodesic. A constant-speed

[1] Various authors call it differently: *shortest path, minimizing geodesic*.

parametrization of a local geodesic by the unit interval $[0, 1]$ is called a *local geodesic path*.

Triangle. For a triple of points $p, q, r \in \mathcal{X}$, a choice of a triple of geodesics $([qr], [rp], [pq])$ will be called a *triangle*; we will use the short notation $[pqr] = ([qr], [rp], [pq])$.

Again, given a triple $p, q, r \in \mathcal{X}$ there may be no triangle $[pqr]$ simply because one of the pairs of these points cannot be joined by a geodesic. Also, many different triangles with these vertices may exist, any of which can be denoted by $[pqr]$. However, if we write $[pqr]$, it means that we have made a choice of such a triangle; that is, we have fixed a choice of the geodesics $[qr]$, $[rp]$, and $[pq]$.

The value

$$|p - q| + |q - r| + |r - p|$$

will be called the *perimeter of the triangle* $[pqr]$.

Hinge. Let $p, x, y \in \mathcal{X}$ be a triple of points such that p is distinct from x and y. A pair of geodesics $([px], [py])$ will be called a *hinge* and will be denoted by $[p\,^x_y] = ([px], [py])$.

Convex set. A set A in a metric space $\mathcal{X}$ is called *convex* if for every two points $p, q \in A$, *every* geodesic $[pq]$ in $\mathcal{X}$ lies in A.

A set $A \subset \mathcal{X}$ is called *locally convex* if every point $a \in A$ admits an open neighborhood $\Omega \ni a$ in $\mathcal{X}$ such that any geodesic lying in Ω and with ends in A lies completely in A.

Note that any open set is locally convex by definition.

1.4 Length spaces

A *curve* is defined as a continuous map from a real interval to a space. If the real interval is $[0, 1]$, then the curve is called a *path*.

1.4.1. Definition. *Let $\mathcal{X}$ be a metric space and $\alpha\colon \mathbb{I} \to \mathcal{X}$ be a curve. We define the* length *of α as*

$$\operatorname{length}\alpha \stackrel{def}{=\!=} \sup_{t_0 \leqslant t_1 \leqslant \ldots \leqslant t_n} \sum_i |\alpha(t_i) - \alpha(t_{i-1})|.$$

Directly from the definition, it follows that if a path $\alpha\colon [0, 1] \to \mathcal{X}$ connects two points x and y (that is, if $\alpha(0) = x$ and $\alpha(1) = y$), then

$$\operatorname{length}\alpha \geqslant |x - y|.$$

Let A be a subset of a metric space $\mathcal{X}$. Given two points $x, y \in A$, consider the value

$$|x - y|_A = \inf_{\alpha}\{\text{length}\,\alpha\},$$

where the infimum is taken for all paths α from x to y in A.[2]

If $|x - y|_A$ takes finite value for each pair $x, y \in A$, then $|x - y|_A$ defines a metric on A; this metric will be called the *induced length metric* on A.

If for any $\varepsilon > 0$ and any pair of points x and y in a metric space $\mathcal{X}$, there is a path α connecting x to y such that

$$\text{length}\,\alpha < |x - y| + \varepsilon,$$

then $\mathcal{X}$ is called a *length space* and the metric on $\mathcal{X}$ is called a *length metric*.

If $f\colon \tilde{\mathcal{X}} \to \mathcal{X}$ is a covering, then a length metric on $\mathcal{X}$ can be lifted to $\tilde{\mathcal{X}}$ by declaring

$$\text{length}_{\tilde{\mathcal{X}}}\,\gamma = \text{length}_{\mathcal{X}}(f \circ \gamma)$$

for any curve γ in $\tilde{\mathcal{X}}$. The space $\tilde{\mathcal{X}}$ with this metric is called the *metric cover* of $\mathcal{X}$.

Note that any geodesic space is a length space. As can be seen from the following example, the converse does not hold.

1.4.2. Example. *Let $\mathcal{X}$ be obtained by gluing a countable collection of disjoint intervals $\{\mathbb{I}_n\}$ of length $1 + \frac{1}{n}$, where for each $\mathbb{I}_n$ the left end is glued to p and the right end to q. Then $\mathcal{X}$ carries a natural complete length metric with respect to which $|p - q| = 1$, but there is no geodesic connecting p to q.*

1.4.3. Exercise. *Give an example of a complete length space for which no pair of distinct points can be joined by a geodesic.*

Let $\mathcal{X}$ be a metric space and $x, y \in \mathcal{X}$.

(i) A point $z \in \mathcal{X}$ is called a *midpoint* between x and y if

$$|x - z| = |y - z| = \tfrac{1}{2} \cdot |x - y|.$$

(ii) Assume $\varepsilon \geqslant 0$. A point $z \in \mathcal{X}$ is called an ε-*midpoint* between x and y if

$$|x - z|, \quad |y - z| \leqslant \tfrac{1}{2} \cdot |x - y| + \varepsilon.$$

Note that a 0-midpoint is the same as a midpoint.

[2] Note that while this notation slightly conflicts with the previously defined notation for distance on a general metric space, we will usually work with ambient length spaces where the meaning will be unambiguous.

1.4.4. Lemma. *Let $\mathcal{X}$ be a complete metric space.*

(a) Assume that for any pair of points $x, y \in \mathcal{X}$ and any $\varepsilon > 0$ there is an ε-midpoint z. Then $\mathcal{X}$ is a length space.
(b) Assume that for any pair of points $x, y \in \mathcal{X}$, there is a midpoint z. Then $\mathcal{X}$ is a geodesic space.

Proof. We first prove (a). Let $x, y \in \mathcal{X}$ be a pair of points.

Set $\varepsilon_n = \frac{\varepsilon}{4^n}$, $\alpha(0) = x$ and $\alpha(1) = y$.

Let $\alpha(\frac{1}{2})$ be an ε_1-midpoint between $\alpha(0)$ and $\alpha(1)$. Further, let $\alpha(\frac{1}{4})$ and $\alpha(\frac{3}{4})$ be ε_2-midpoints between the pairs $(\alpha(0), \alpha(\frac{1}{2}))$ and $(\alpha(\frac{1}{2}), \alpha(1))$, respectively. Applying the above procedure recursively, on the n-th step we define $\alpha(\frac{k}{2^n})$, for every odd integer k such that $0 < \frac{k}{2^n} < 1$, as an ε_n-midpoint between the already defined $\alpha(\frac{k-1}{2^n})$ and $\alpha(\frac{k+1}{2^n})$.

In this way we define $\alpha(t)$ for $t \in W$, where W denotes the set of dyadic rationals in $[0, 1]$. Since $\mathcal{X}$ is complete, the map α can be extended continuously to $[0, 1]$. Moreover,

$$\text{❶} \qquad \begin{aligned} \operatorname{length} \alpha &\leqslant |x - y| + \sum_{n=1}^{\infty} 2^{n-1} \cdot \varepsilon_n \leqslant \\ &\leqslant |x - y| + \tfrac{\varepsilon}{2}. \end{aligned}$$

Since $\varepsilon > 0$ is arbitrary, we get (a).

To prove (b), one should repeat the same argument taking midpoints instead of ε_n-midpoints. In this case ❶ holds for $\varepsilon_n = \varepsilon = 0$. □

Since in a compact space a sequence of $\frac{1}{n}$-midpoints z_n contains a convergent subsequence, Lemma 1.4.4 immediately implies

1.4.5. Proposition. *A proper length space is geodesic.*

1.4.6. Hopf–Rinow theorem. *Any complete, locally compact length space is proper.*

Proof. Let $\mathcal{X}$ be a locally compact length space. Given $x \in \mathcal{X}$, denote by $\rho(x)$ the supremum of all $R > 0$ such that the closed ball $\overline{\mathrm{B}}[x, R]$ is compact. Since $\mathcal{X}$ is locally compact,

$$\text{❷} \qquad \rho(x) > 0 \ \text{ for any } \ x \in \mathcal{X}.$$

It is sufficient to show that $\rho(x) = \infty$ for some (and therefore any) point $x \in \mathcal{X}$.

Assume the contrary; that is, $\rho(x) < \infty$. We claim that

❸ $B = \overline{\mathrm{B}}[x, \rho(x)]$ *is compact for any* x.

Indeed, $\mathcal{X}$ is a length space; therefore for any $\varepsilon > 0$, the set $\overline{\mathrm{B}}[x, \rho(x) - \varepsilon]$ is a compact ε-net in B. Since B is closed and hence complete, it must be compact. △

Next we claim that

❹ $|\rho(x) - \rho(y)| \leqslant |x - y|_{\mathcal{X}}$ *for any* $x, y \in \mathcal{X}$; *in particular* $\rho\colon \mathcal{X} \to \mathbb{R}$ *is a continuous function.*

Indeed, assume the contrary; that is, $\rho(x) + |x - y| < \rho(y)$ for some $x, y \in \mathcal{X}$. Then $\overline{\mathrm{B}}[x, \rho(x) + \varepsilon]$ is a closed subset of $\overline{\mathrm{B}}[y, \rho(y)]$ for some $\varepsilon > 0$. Then compactness of $\overline{\mathrm{B}}[y, \rho(y)]$ implies compactness of $\overline{\mathrm{B}}[x, \rho(x) + \varepsilon]$, a contradiction. △

Set $\varepsilon = \min\{\, \rho(y) : y \in B \,\}$; the minimum is defined since B is compact. From ❷, we have $\varepsilon > 0$.

Choose a finite $\frac{\varepsilon}{10}$-net $\{a_1, a_2, \dots, a_n\}$ in B. The union W of the closed balls $\overline{\mathrm{B}}[a_i, \varepsilon]$ is compact. Clearly $\overline{\mathrm{B}}[x, \rho(x) + \frac{\varepsilon}{10}] \subset W$. Therefore $\overline{\mathrm{B}}[x, \rho(x) + \frac{\varepsilon}{10}]$ is compact, a contradiction. □

1.4.7. Exercise. *Construct a geodesic space that is locally compact, but whose completion is neither geodesic nor locally compact.*

1.5 Model angles and triangles

Let $\mathcal{X}$ be a metric space and $p, q, r \in \mathcal{X}$. Let us define the *model triangle* $[\tilde{p}\tilde{q}\tilde{r}]$ (briefly, $[\tilde{p}\tilde{q}\tilde{r}] = \tilde{\triangle}(pqr)_{\mathbb{E}^2}$) to be a triangle in the plane $\mathbb{E}^2$ with the same side lengths; that is,

$$|\tilde{p} - \tilde{q}| = |p - q|, \quad |\tilde{q} - \tilde{r}| = |q - r|, \quad |\tilde{r} - \tilde{p}| = |r - p|.$$

In the same way we can define the *hyperbolic* and the *spherical model triangles* $\tilde{\triangle}(pqr)_{\mathbb{H}^2}$, $\tilde{\triangle}(pqr)_{\mathbb{S}^2}$ in the hyperbolic plane $\mathbb{H}^2$ and the unit sphere $\mathbb{S}^2$. In the latter case the model triangle is said to be defined if in addition

$$|p - q| + |q - r| + |r - p| < 2 \cdot \pi.$$

In this case the model triangle again exists and is unique up to an isometry of $\mathbb{S}^2$.

If $[\tilde{p}\tilde{q}\tilde{r}] = \tilde{\triangle}(pqr)_{\mathbb{E}^2}$ and $|p - q|, |p - r| > 0$, the angle measure of $[\tilde{p}\tilde{q}\tilde{r}]$ at $\tilde{p}$ will be called the *model angle* of the triple p, q, r and will be denoted by $\tilde{\measuredangle}(p\,{}^q_r)_{\mathbb{E}^2}$. In the same way we define $\tilde{\measuredangle}(p\,{}^q_r)_{\mathbb{H}^2}$ and $\tilde{\measuredangle}(p\,{}^q_r)_{\mathbb{S}^2}$; in the latter case we assume in addition that the model triangle $\tilde{\triangle}(pqr)_{\mathbb{S}^2}$ is defined.

We may use the notation $\tilde{\measuredangle}(p\,{}^q_r)$ if it is evident which of the model spaces $\mathbb{H}^2$, $\mathbb{E}^2$, or $\mathbb{S}^2$ is meant.

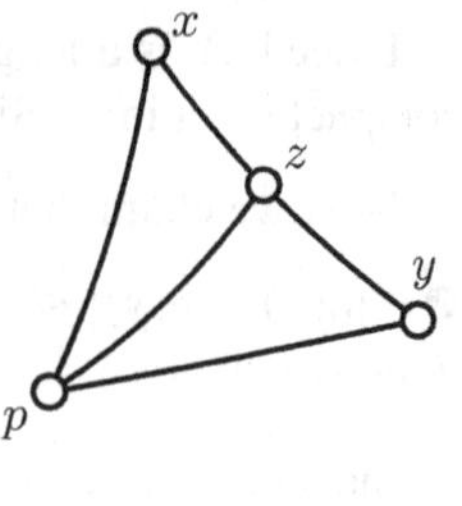

1.5.1. Alexandrov's lemma. *Let p, x, y, z be distinct points in a metric space such that $z \in]xy[$. Then the following expressions for the Euclidean model angles have the same sign:*

(a) $\tilde\measuredangle(x\,{}^{p}_{y}) - \tilde\measuredangle(x\,{}^{p}_{z})$,
(b) $\tilde\measuredangle(z\,{}^{p}_{x}) + \tilde\measuredangle(z\,{}^{p}_{y}) - \pi$.

Moreover,

$$\tilde\measuredangle(p\,{}^{x}_{y}) \geqslant \tilde\measuredangle(p\,{}^{x}_{z}) + \tilde\measuredangle(p\,{}^{z}_{y}),$$

with equality if and only if the expressions in (a) and (b) vanish.

The same holds for the hyperbolic and spherical model angles, but in the latter case one has to assume in addition that

$$|p - z| + |p - y| + |x - y| < 2 \cdot \pi.$$

Proof. Consider the model triangle $[\tilde x \tilde p \tilde z] = \tilde\triangle(xpz)$. Take a point $\tilde y$ on the extension of $[\tilde x \tilde z]$ beyond $\tilde z$ so that $|\tilde x - \tilde y| = |x - y|$ (and therefore $|\tilde x - \tilde z| = |x - z|$).

Since increasing the opposite side in a plane triangle increases the corresponding angle, the following expressions have the same sign:

(i) $\measuredangle[\tilde x\,{}^{\tilde p}_{\tilde y}] - \tilde\measuredangle(x\,{}^{p}_{y})$,
(ii) $|\tilde p - \tilde y| - |p - y|$,
(iii) $\measuredangle[\tilde z\,{}^{\tilde p}_{\tilde y}] - \tilde\measuredangle(z\,{}^{p}_{y})$.

Since

$$\measuredangle[\tilde x\,{}^{\tilde p}_{\tilde y}] = \measuredangle[\tilde x\,{}^{\tilde p}_{\tilde z}] = \tilde\measuredangle(x\,{}^{p}_{z})$$

and

$$\measuredangle[\tilde z\,{}^{\tilde p}_{\tilde y}] = \pi - \measuredangle[\tilde z\,{}^{\tilde x}_{\tilde p}] = \pi - \tilde\measuredangle(z\,{}^{x}_{p}),$$

the first statement follows.

For the second statement, construct a model triangle $[\tilde p \tilde z \tilde y'] = \tilde\triangle(pzy)_{\mathbb{E}^2}$ on the opposite side of $[\tilde p \tilde z]$ from $[\tilde x \tilde p \tilde z]$. Note that

$$\begin{aligned} |\tilde x - \tilde y'| &\leqslant |\tilde x - \tilde z| + |\tilde z - \tilde y'| = \\ &= |x - z| + |z - y| = \\ &= |x - y|. \end{aligned}$$

Therefore

$$\begin{aligned}\tilde\measuredangle(p\,{}^{x}_{z}) + \tilde\measuredangle(p\,{}^{z}_{y}) &= \measuredangle[\tilde p\,{}^{\tilde x}_{\tilde z}] + \measuredangle[\tilde p\,{}^{\tilde z}_{\tilde y'}] = \\ &= \measuredangle[\tilde p\,{}^{\tilde x}_{\tilde y'}] \leqslant \\ &\leqslant \tilde\measuredangle(p\,{}^{x}_{y}).\end{aligned}$$

Equality holds if and only if $|\tilde x - \tilde y'| = |x - y|$, as required. □

1.6 Angles and the first variation

Given a hinge $[p\,{}^{x}_{y}]$, we define its *angle* as the limit

$$\measuredangle[p\,{}^{x}_{y}] \stackrel{def}{=\!=} \lim_{\bar x, \bar y \to p} \tilde\measuredangle(p\,{}^{\bar x}_{\bar y})_{\mathbb{E}^2}, \qquad ❶$$

where $\bar x \in\,]px]$ and $\bar y \in\,]py]$. (The angle $\measuredangle[p\,{}^{x}_{y}]$ is defined if the limit exists.)

The value under the limit can be calculated from the cosine law:

$$\cos \tilde\measuredangle(p\,{}^{x}_{y})_{\mathbb{E}^2} = \frac{|p-x|^2 + |p-y|^2 - |x-y|^2}{2\cdot|p-x|\cdot|p-y|}.$$

The following lemma implies that in ❶, one can use $\tilde\measuredangle(p\,{}^{\bar x}_{\bar y})_{\mathbb{S}^2}$ or $\tilde\measuredangle(p\,{}^{\bar x}_{\bar y})_{\mathbb{H}^2}$ instead of $\tilde\measuredangle(p\,{}^{\bar x}_{\bar y})_{\mathbb{E}^2}$.

1.6.1. Lemma. *For any three points p, x, y in a metric space the following inequalities*

$$\begin{aligned}|\tilde\measuredangle(p\,{}^{x}_{y})_{\mathbb{S}^2} - \tilde\measuredangle(p\,{}^{x}_{y})_{\mathbb{E}^2}| &\leqslant |p-x|\cdot|p-y|, \\ |\tilde\measuredangle(p\,{}^{x}_{y})_{\mathbb{H}^2} - \tilde\measuredangle(p\,{}^{x}_{y})_{\mathbb{E}^2}| &\leqslant |p-x|\cdot|p-y|\end{aligned} \qquad ❷$$

hold whenever the left-hand side is defined.

Proof. Note that

$$\tilde\measuredangle(p\,{}^{x}_{y})_{\mathbb{H}^2} \leqslant \tilde\measuredangle(p\,{}^{x}_{y})_{\mathbb{E}^2} \leqslant \tilde\measuredangle(p\,{}^{x}_{y})_{\mathbb{S}^2}.$$

Therefore

$$\begin{aligned}0 &\leqslant \tilde\measuredangle(p\,{}^{x}_{y})_{\mathbb{S}^2} - \tilde\measuredangle(p\,{}^{x}_{y})_{\mathbb{H}^2} \leqslant \\ &\leqslant \tilde\measuredangle(p\,{}^{x}_{y})_{\mathbb{S}^2} + \tilde\measuredangle(x\,{}^{p}_{y})_{\mathbb{S}^2} + \tilde\measuredangle(y\,{}^{p}_{x})_{\mathbb{S}^2} - \tilde\measuredangle(p\,{}^{x}_{y})_{\mathbb{H}^2} - \tilde\measuredangle(x\,{}^{p}_{y})_{\mathbb{H}^2} - \tilde\measuredangle(y\,{}^{p}_{x})_{\mathbb{H}^2} = \\ &= \operatorname{area}\tilde\triangle(pxy)_{\mathbb{S}^2} + \operatorname{area}\tilde\triangle(pxy)_{\mathbb{H}^2}.\end{aligned}$$

The inequality ❷ follows since

$$\begin{aligned} 0 &\leqslant \operatorname{area}\tilde{\triangle}(pxy)_{\mathbb{H}^2} \leqslant \\ &\leqslant \operatorname{area}\tilde{\triangle}(pxy)_{\mathbb{S}^2} \leqslant \\ &\leqslant |p-x|\cdot|p-y|. \end{aligned}$$

□

1.6.2. Triangle inequality for angles. *Let* $[px^1]$, $[px^2]$, *and* $[px^3]$ *be three geodesics in a metric space. If all the angles* $\alpha^{ij} = \measuredangle[p\,{}^{x^i}_{x^j}]$ *are defined, then they satisfy the triangle inequality:*

$$\alpha^{13} \leqslant \alpha^{12} + \alpha^{23}.$$

Proof. Since $\alpha^{13} \leqslant \pi$, we may assume that $\alpha^{12} + \alpha^{23} < \pi$.

Set $\gamma^i = \operatorname{geod}_{[px^i]}$. Given any $\varepsilon > 0$, for all sufficiently small $t, \tau, s \in \mathbb{R}_+$ we have

$$\begin{aligned} |\gamma^1(t) - \gamma^3(\tau)| \leqslant& |\gamma^1(t) - \gamma^2(s)| + |\gamma^2(s) - \gamma^3(\tau)| < \\ &< \sqrt{t^2 + s^2 - 2\cdot t\cdot s\cdot\cos(\alpha^{12}+\varepsilon)} + \\ &+ \sqrt{s^2 + \tau^2 - 2\cdot s\cdot\tau\cdot\cos(\alpha^{23}+\varepsilon)} \leqslant \end{aligned}$$

Below we define $s(t,\tau)$ so that for $s = s(t,\tau)$, this chain of inequalities can be continued as follows:

$$\leqslant \sqrt{t^2 + \tau^2 - 2\cdot t\cdot\tau\cdot\cos(\alpha^{12}+\alpha^{23}+2\cdot\varepsilon)}.$$

Thus for any $\varepsilon > 0$,

$$\alpha^{13} \leqslant \alpha^{12} + \alpha^{23} + 2\cdot\varepsilon.$$

Hence the result.

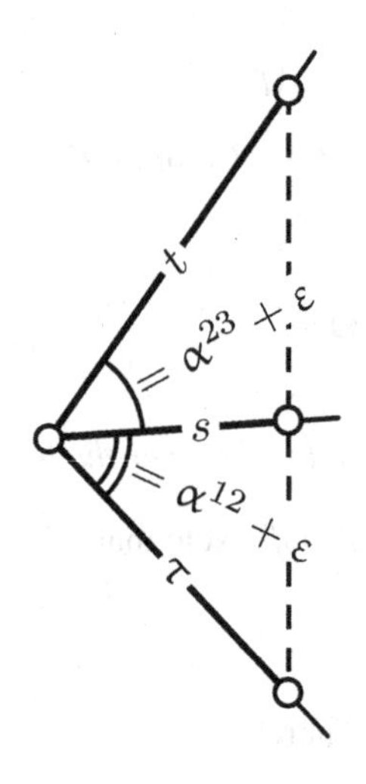

To define $s(t,\tau)$, consider three rays $\tilde{\gamma}^1, \tilde{\gamma}^2, \tilde{\gamma}^3$ on a Euclidean plane starting at one point, such that $\measuredangle(\tilde{\gamma}^1, \tilde{\gamma}^2) = \alpha^{12} + \varepsilon$, $\measuredangle(\tilde{\gamma}^2, \tilde{\gamma}^3) = \alpha^{23} + \varepsilon$ and $\measuredangle(\tilde{\gamma}^1, \tilde{\gamma}^3) = \alpha^{12} + \alpha^{23} + 2\cdot\varepsilon$. We parametrize each ray by the distance from the starting point. Given two positive numbers $t, \tau \in \mathbb{R}_+$, let $s = s(t,\tau)$ be the number such that $\tilde\gamma^2(s) \in [\tilde\gamma^1(t)\ \tilde\gamma^3(\tau)]$. Clearly $s \leqslant \max\{t,\tau\}$, so t,τ,s may be taken sufficiently small. □

1.6.3. Exercise. *Prove that the sum of adjacent angles is at least* π.

More precisely, let $\mathcal{X}$ *be a complete length space and* $p, x, y, z \in \mathcal{X}$. *If* $p \in\,]xy[$, *then*

$$\measuredangle[p\,{}^x_z] + \measuredangle[p\,{}^y_z] \geqslant \pi$$

whenever each angle on the left-hand side is defined.

1.6.4. First variation inequality. *Assume that for a hinge* $[q\,{}^p_x]$ *the angle* $\alpha = \measuredangle[q\,{}^p_x]$ *is defined. Then*

$$|p - \mathrm{geod}_{[qx]}(t)| \leqslant |q - p| - t\cdot\cos\alpha + o(t).$$

Proof. Take a sufficiently small $\varepsilon > 0$. For all sufficiently small $t > 0$, we have

$$\begin{aligned}|\mathrm{geod}_{[qp]}(t/\varepsilon) - \mathrm{geod}_{[qx]}(t)| &\leqslant \tfrac{t}{\varepsilon}\cdot\sqrt{1+\varepsilon^2 - 2\cdot\varepsilon\cdot\cos\alpha} + o(t) \leqslant \\ &\leqslant \tfrac{t}{\varepsilon} - t\cdot\cos\alpha + t\cdot\varepsilon.\end{aligned}$$

Applying the triangle inequality, we get

$$\begin{aligned}|p - \mathrm{geod}_{[qx]}(t)| &\leqslant |p - \mathrm{geod}_{[qp]}(t/\varepsilon)| + |\mathrm{geod}_{[qp]}(t/\varepsilon) - \mathrm{geod}_{[qx]}(t)| \leqslant \\ &\leqslant |p-q| - t\cdot\cos\alpha + t\cdot\varepsilon\end{aligned}$$

for any fixed $\varepsilon > 0$ and all sufficiently small t. Hence the result. □

1.7 Space of directions and tangent space

Let $\mathcal{X}$ be a metric space with defined angles for all hinges. Fix a point $p \in \mathcal{X}$.

Consider the set $\mathfrak{S}_p$ of all nontrivial geodesics that start at p. By 1.6.2, the triangle inequality holds for $\measuredangle$ on $\mathfrak{S}_p$, so $(\mathfrak{S}_p, \measuredangle)$ forms a *pseudometric space*; that is, $\measuredangle$ satisfies all the conditions of a metric on $\mathfrak{S}_p$, except that the angle between distinct geodesics might vanish.

The metric space corresponding to $(\mathfrak{S}_p, \measuredangle)$ is called the *space of geodesic directions* at p, denoted by Σ'_p or $\Sigma'_p\mathcal{X}$. Elements of Σ'_p are called *geodesic directions* at p. Each geodesic direction is formed by an equivalence class of geodesics in $\mathfrak{S}_p$ for the equivalence relation

$$[px] \sim [py] \iff \measuredangle[p\,{}^x_y] = 0.$$

The completion of Σ'_p is called the *space of directions* at p and is denoted by Σ_p or $\Sigma_p\mathcal{X}$. Elements of Σ_p are called *directions* at p.

The Euclidean cone Cone Σ_p over the space of directions Σ_p is called the *tangent space* at p and is denoted by T_p or $\mathrm{T}_p\mathcal{X}$.

The tangent space T_p could also be defined directly, without introducing the space of directions. To do so, consider the set $\mathfrak{T}_p$ of all geodesics with constant-speed parametrizations starting at p. Given $\alpha, \beta \in \mathfrak{T}_p$, set

$$❶ \qquad |\alpha - \beta|_{\mathfrak{T}_p} = \lim_{\varepsilon \to 0} \frac{|\alpha(\varepsilon) - \beta(\varepsilon)|_{\mathcal{X}}}{\varepsilon}$$

Since the angles in $\mathcal{X}$ are defined, ❶ defines a pseudometric on $\mathfrak{T}_p$.

The corresponding metric space admits a natural isometric identification with the cone $\mathrm{T}'_p = \mathrm{Cone}\,\Sigma'_p$. The elements of T'_p are equivalence classes for the relation

$$\alpha \sim \beta \iff |\alpha(t) - \beta(t)|_{\mathcal{X}} = o(t).$$

The completion of T'_p is therefore naturally isometric to T_p.

Elements of T_p will be called *tangent vectors* at p, regardless of the fact that T_p is only a metric cone and need not be a vector space. Elements of T'_p will be called *geodesic tangent vectors* at p.

1.8 Hausdorff convergence

It seems that *Hausdorff convergence* was first introduced by Felix Hausdorff in [42], and a couple of years later an equivalent definition was given by Wilhelm Blaschke in [16]. A refinement of this definition was introduced by Zdeněk Frolík in [35], and later rediscovered by Robert Wijsman in [68]. However, this refinement takes an intermediate place between the original Hausdorff convergence and *closed convergence*, also introduced by Hausdorff in [42]. For this reason we call it Hausdorff convergence instead of *Hausdorff–Blascke–Frolík–Wijsman convergence*.

Let $\mathcal{X}$ be a metric space and $A \subset \mathcal{X}$. We will denote by $\mathrm{dist}_A(x)$ the distance from A to a point x in $\mathcal{X}$; that is,

$$\mathrm{dist}_A(x) \stackrel{def}{=\!=} \inf\{\, |a - x|_{\mathcal{X}} : a \in A \,\}.$$

1.8.1. Definition of Hausdorff convergence. *Given a sequence of closed sets $(A_n)_{n=1}^{\infty}$ in a metric space $\mathcal{X}$, a closed set $A_\infty \subset \mathcal{X}$ is called the* Hausdorff limit *of $(A_n)_{n=1}^{\infty}$, briefly $A_n \to A_\infty$, if*

$$\operatorname{dist}_{A_n}(x) \to \operatorname{dist}_{A_\infty}(x) \ \ as \ \ n \to \infty$$

for every $x \in \mathcal{X}$.

In this case, the sequence of closed sets $(A_n)_{n=1}^{\infty}$ is said to converge in the sense of Hausdorff.

Example. Let D_n be the disk in the coordinate plane with center $(0, n)$ and radius n. Then D_n converges to the upper half-plane as $n \to \infty$.

1.8.2. Exercise. *Let $A_n \to A_\infty$ as in Definition 1.8.1.*

Show that A_∞ is the set of all points p such that $p_n \to p$ for some sequence of points $p_n \in A_n$.

Does the converse hold? That is, suppose $(A_n)_{n=1}^{\infty}$, A_∞ are closed sets such that A_∞ is the set of all points p such that $p_n \to p$ for some sequence of points $p_n \in A_n$. Does this imply that $A_n \to A_\infty$?

1.8.3. First selection theorem. *Let $\mathcal{X}$ be a proper metric space and $(A_n)_{n=1}^{\infty}$ be a sequence of closed sets in $\mathcal{X}$. Assume that for some (and therefore any) point $x \in \mathcal{X}$, the sequence $\operatorname{dist}_{A_n}(x)$ is bounded. Then the sequence $(A_n)_{n=1}^{\infty}$ has a convergent subsequence in the sense of Hausdorff.*

Proof. Since X is proper, there is a countable dense set $\{x_1, x_2, \dots\}$ in $\mathcal{X}$. Note that the sequence $d_n = \operatorname{dist}_{A_n}(x_k)$ is bounded for each k. Therefore, passing to a subsequence of $(A_n)_{n=1}^{\infty}$, we can assume that $\operatorname{dist}_{A_n}(x_k)$ converges as $n \to \infty$ for any fixed k.

Note that for each n, the function $\operatorname{dist}_{A_n} \colon \mathcal{X} \to \mathbb{R}$ is 1-Lipschitz and nonnegative. Therefore the sequence $\operatorname{dist}_{A_n}$ converges pointwise to a 1-Lipschitz nonnegative function $f \colon \mathcal{X} \to \mathbb{R}$.

Set $A_\infty = f^{-1}(0)$. Let us show that

$$\operatorname{dist}_{A_\infty}(y) \leqslant f(y)$$

for any y.

Assume the contrary; that is,

$$f(z) < R < \operatorname{dist}_{A_\infty}(z)$$

for some $z \in \mathcal{X}$ and $R > 0$. Then for any sufficiently large n, there is a point $z_n \in A_n$ such that $|x - z_n| \leqslant R$. Since $\mathcal{X}$ is proper, we can pass to a partial limit z_∞ of z_n as $n \to \infty$.

It is clear that $f(z_\infty) = 0$; that is, $z_\infty \in A_\infty$. (Note that this implies that $A_\infty \neq \varnothing$). On the other hand,

$$\operatorname{dist}_{A_\infty}(y) \leqslant |z_\infty - y| \leqslant R < \operatorname{dist}_{A_\infty}(y),$$

a contradiction.

On the other hand, since f is 1-Lipschitz, $\operatorname{dist}_{A_\infty}(y) \geqslant f(y)$. Therefore

$$\operatorname{dist}_{A_\infty}(y) = f(y)$$

for any $y \in \mathcal{X}$. Hence the result. □

1.9 Gromov–Hausdorff convergence

1.9.1. Definition. *Let* $\{\mathcal{X}_\alpha : \alpha \in \mathcal{A}\}$ *be a collection of metric spaces. A metric* ρ *on the disjoint union*

$$\boldsymbol{X} = \bigsqcup_{\alpha\in\mathcal{A}} \mathcal{X}_\alpha$$

is called a compatible metric *if the restriction of* ρ *to every* $\mathcal{X}_\alpha$ *coincides with the original metric on* $\mathcal{X}_\alpha$.

1.9.2. Definition. *Let* $\mathcal{X}_1, \mathcal{X}_2, \ldots$ *and* $\mathcal{X}_\infty$ *be proper metric spaces and* ρ *be a compatible metric on their disjoint union* $\boldsymbol{X}$. *Assume that* $\mathcal{X}_n$ *is an open set in* $(\boldsymbol{X}, \rho)$ *for each* $n \neq \infty$, *and* $\mathcal{X}_n \to \mathcal{X}_\infty$ *in* $(\boldsymbol{X}, \rho)$ *as* $n \to \infty$ *in the sense of Hausdorff (see Definition 1.8.1).*

Then we say ρ *defines a* convergence[3] in the sense of Gromov–Hausdorff, *and write* $\mathcal{X}_n \to \mathcal{X}_\infty$ *or* $\mathcal{X}_n \xrightarrow{\rho} \mathcal{X}_\infty$. *The space* $\mathcal{X}_\infty$ *is called the* limit space *of the sequence* $(\mathcal{X}_n)$ along ρ.

Usually Gromov–Hausdorff convergence is defined differently. We prefer this definition since it induces convergence for a sequence of points $x_n \in \mathcal{X}_n$ (Exercise 1.8.2), as well as weak convergence of measures μ_n on $\mathcal{X}_n$, and so on, corresponding to convergence in the ambient space $(\boldsymbol{X}, \rho)$.

Once we write $\mathcal{X}_n \to \mathcal{X}_\infty$, we mean that we have made a choice of convergence. Note that for a fixed sequence of metric spaces $(\mathcal{X}_n)$, it might be possible to construct different Gromov–Hausdorff convergences, say $\mathcal{X}_n \xrightarrow{\rho} \mathcal{X}_\infty$ and $\mathcal{X}_n \xrightarrow{\rho'} \mathcal{X}'_\infty$, whose limit spaces $\mathcal{X}_\infty$ and $\mathcal{X}'_\infty$ need not be isometric to each other.

For example, for the constant sequence $\mathcal{X}_n \overset{iso}{=\!=} \mathbb{R}_{\geqslant 0}$, there is a convergence with limit $\mathcal{X}_\infty \overset{iso}{=\!=} \mathbb{R}_{\geqslant 0}$; guess the metric ρ from the diagram.

---

[3]Formally speaking, convergence is the topology induced by ρ on $\boldsymbol{X}$.

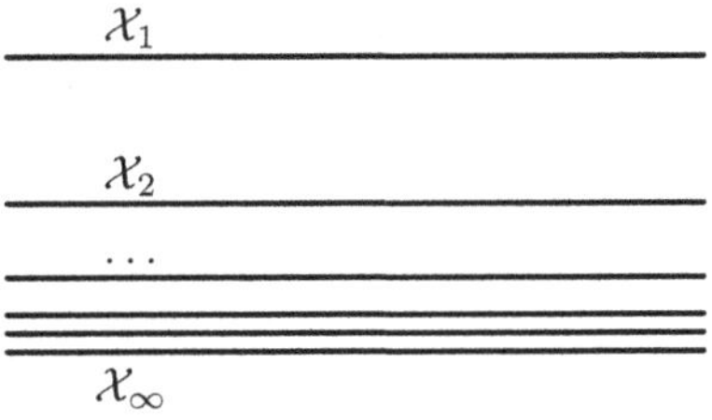

$\mathcal{X}_1$

$\mathcal{X}_2$

...

$\mathcal{X}_\infty$

For another metric ρ'—also guess it from the diagram—the limit space $\mathcal{X}'_\infty$ is isometric to the real line.

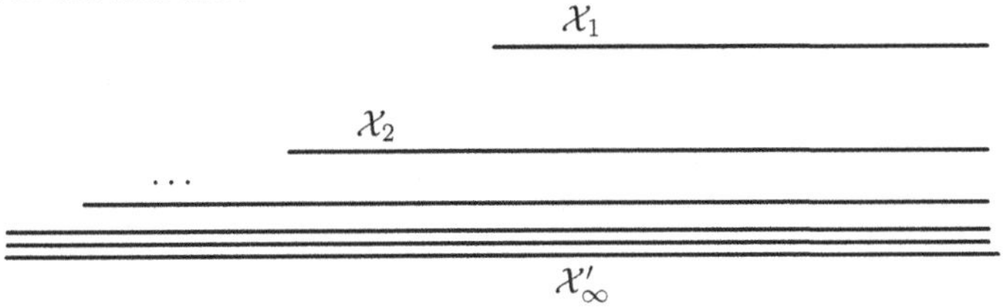

$\mathcal{X}_1$

$\mathcal{X}_2$

...

$\mathcal{X}'_\infty$

1.9.3. Second selection theorem. *Let $\mathcal{X}_n$ be a sequence of proper metric spaces with marked points $x_n \in \mathcal{X}_n$. Assume that for any fixed $R, \varepsilon > 0$, there is $N = N(R, \varepsilon) \in \mathbb{N}$ such that for each n the ball $\overline{\mathrm{B}}[x_n, R]_{\mathcal{X}_n}$ admits a finite ε-net with at most N points. Then there is a subsequence of $\mathcal{X}_n$ admitting a Gromov–Hausdorff convergence such that the sequence of marked points $x_n \in \mathcal{X}_n$ converges.*

Proof. From the main assumption in the theorem, in each space $\mathcal{X}_n$ there is a sequence of points $z_{i,n} \in \mathcal{X}_n$ such that the following condition holds for a fixed sequence of integers $M_1 < M_2 < \dots$

- $|z_{i,n} - x_n|_{\mathcal{X}_n} \leqslant k + 1$ if $i \leqslant M_k$,
- the points $z_{1,n}, \dots, z_{M_k,n}$ form an $\frac{1}{k}$-net in $\overline{\mathrm{B}}[x_n, k]_{\mathcal{X}_n}$.

Passing to a subsequence, we can assume that the sequence

$$\ell_n = |z_{i,n} - z_{j,n}|_{\mathcal{X}_n}$$

converges for any i and j.

Consider a countable set of points $\mathcal{W} = \{w_1, w_2, \dots\}$ equipped with the pseudometric defined by

$$|w_i - w_j|_{\mathcal{W}} = \lim_{n \to \infty} |z_{i,n} - z_{j,n}|_{\mathcal{X}_n}.$$

Let $\hat{\mathcal{W}}$ be the metric space corresponding to $\mathcal{W}$; that is, points in $\hat{\mathcal{W}}$ are equivalence classes in $\mathcal{W}$ for the relation $\sim$, where $w_i \sim w_j$ if and only if $|w_i - w_j|_{\mathcal{W}} = 0$, and where

$$|[w_i] - [w_j]|_{\hat{\mathcal{W}}} \stackrel{def}{=\!=} |w_i - w_j|_{\mathcal{W}}.$$

Denote by $\mathcal{X}_\infty$ the completion of $\hat{\mathcal{W}}$.

It remains to show that there is a Gromov–Hausdorff convergence $\mathcal{X}_n \to \mathcal{X}_\infty$ such that the sequence $x_n \in \mathcal{X}_n$ converges. To prove this, we need to construct a metric ρ on the disjoint union of

$$X = \mathcal{X}_\infty \sqcup \mathcal{X}_1 \sqcup \mathcal{X}_2 \sqcup \dots$$

satisfying definitions 1.9.1 and 1.9.2. The metric ρ can be constructed as the maximal compatible metric such that

$$\rho(z_{i,n}, w_i) \leqslant \tfrac{1}{m}$$

for any $n \geqslant N_m$ and $i < I_m$ for a suitable choice of two sequences (I_m) and (N_m) with $I_1 = N_1 = 1$. $\square$

1.9.4. Exercise. *Let $\mathcal{X}_n$ be a sequence of metric spaces that admits two convergences $\mathcal{X}_n \xrightarrow{\rho} \mathcal{X}_\infty$ and $\mathcal{X}_n \xrightarrow{\rho'} \mathcal{X}'_\infty$.*

(a) If $\mathcal{X}_\infty$ is compact, then $\mathcal{X}_\infty \overset{iso}{=\!=} \mathcal{X}'_\infty$.
(b) If $\mathcal{X}_\infty$ is proper and there is a sequence of points $x_n \in \mathcal{X}_n$ that converges in both convergences, then $\mathcal{X}_\infty \overset{iso}{=\!=} \mathcal{X}'_\infty$.

Chapter 2
Gluing theorem and billiards

In this chapter we define CAT(κ) spaces and give the first application, to billiards.

Here "CAT" is an acronym for Cartan, Alexandrov, and Toponogov. It was coined by Mikhael Gromov in 1987. Originally, Alexandrov called these spaces "$\mathfrak{R}_\kappa$ domain"; this term is still in use.

Riemannian manifolds with nonpositive sectional curvature provide a motivating example. Specifically, a Riemannian manifold has nonpositive sectional curvature if and only if each point admits a CAT(0) neighborhood.

2.1 The 4-point condition

Given a quadruple of points p, q, x, y in a metric space $\mathcal{X}$, consider two model triangles in the plane $[\tilde{p}\tilde{x}\tilde{y}] = \tilde{\triangle}(pxy)_{\mathbb{E}^2}$ and $[\tilde{q}\tilde{x}\tilde{y}] = \tilde{\triangle}(qxy)_{\mathbb{E}^2}$ with common side $[\tilde{x}\tilde{y}]$.

If the inequality

$$|p - q|_{\mathcal{X}} \leqslant |\tilde{p} - \tilde{z}|_{\mathbb{E}^2} + |\tilde{z} - \tilde{q}|_{\mathbb{E}^2}$$

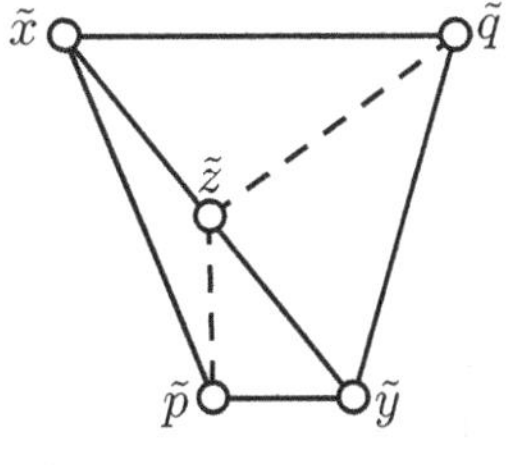

holds for any point $\tilde{z} \in [\tilde{x}\tilde{y}]$, then we say that the quadruple p, q, x, y *satisfies* CAT(0) *comparison*.

If we do the same for spherical model triangles $[\tilde{p}\tilde{x}\tilde{y}] = \tilde{\triangle}(pxy)_{\mathbb{S}^2}$ and $[\tilde{q}\tilde{x}\tilde{y}] = \tilde{\triangle}(qxy)_{\mathbb{S}^2}$, then we arrive at the definition of CAT(1) comparison. If one of the spherical model triangles is undefined,[1] then it is assumed that CAT(1) comparison automatically holds for this quadruple.

[1] That is, if

$$|p - x| + |p - y| + |x - y| \geqslant 2 \cdot \pi \quad \text{or} \quad |q - x| + |q - y| + |x - y| \geqslant 2 \cdot \pi.$$

S. Alexander et al., *An Invitation to Alexandrov Geometry*,
SpringerBriefs in Mathematics, https://doi.org/10.1007/978-3-030-05312-3_2

We can do the same for the model plane of curvature κ; that is, a sphere if $\kappa > 0$, Euclidean plane if $\kappa = 0$ and Lobachevsky plane if $\kappa < 0$. In this case we arrive at the definition of CAT(κ) comparison. However in these notes we will mostly consider CAT(0) comparison and occasionally CAT(1) comparison; so, if you see CAT(κ), you can assume that κ is 0 or 1.

If all quadruples in a metric space $\mathcal{X}$ satisfy CAT(κ) comparison, then we say that the space $\mathcal{X}$ is CAT(κ). (Note that CAT(κ) is an adjective.)

In order to check CAT(κ) comparison, it is sufficient to know the 6 distances between all pairs of points in the quadruple. This observation implies the following.

2.1.1. Proposition. *Any Gromov–Hausdorff limit of a sequence of* CAT(κ) *spaces is* CAT(κ).

In the proposition above, it does not matter which definition of convergence for metric spaces you use, as long as any quadruple of points in the limit space can be arbitrarily well approximated by quadruples in the sequence of metric spaces.

2.1.2. Exercise. *Let $\mathcal{V}$ be a metric space and $\mathcal{U} = \operatorname{Cone}\mathcal{V}$. Show that $\mathcal{U}$ is* CAT(0) *if and only if $\mathcal{V}$ is* CAT(1).

Analogously, if $\mathcal{U} = \operatorname{Susp}\mathcal{V}$, then $\mathcal{U}$ is CAT(1) *if and only if $\mathcal{V}$ is* CAT(1).

The cone and suspension constructions are defined in Section 1.2.

The following exercise is a bit simpler, but can be proved in essentially the same way.

2.1.3. Exercise. *Assume $\mathcal{U}$ and $\mathcal{V}$ are* CAT(0) *spaces. Show that the product space $\mathcal{U} \times \mathcal{V}$ is* CAT(0).

2.1.4. Exercise. *Show that any complete length* CAT(0) *space is geodesic.*

2.2 Thin triangles

The inheritance lemma 2.2.9 proved below plays a central role in the theory of CAT(κ) spaces. It will lead to two fundamental constructions: patchwork globalization (3.3.2) and Reshetnyak gluing (2.3.1), which in turn are used to prove the globalization theorem (3.3.1).

Recall that a *triangle* $[x^1x^2x^3]$ in a space $\mathcal{X}$ is a triple of minimizing geodesics $[x^1x^2]$, $[x^2x^3]$, and $[x^3x^1]$. Consider the model triangle $[\tilde{x}^1\tilde{x}^2\tilde{x}^3] = \tilde{\triangle}(x^1x^2x^3)_{\mathbb{E}^2}$ in the Euclidean plane. The *natural map* $[\tilde{x}^1\tilde{x}^2\tilde{x}^3] \to [x^1x^2x^3]$ sends a point $\tilde{z} \in [\tilde{x}^i\tilde{x}^j]$ to the corresponding point $z \in [x^ix^j]$; that is, z is the point such that $|\tilde{x}^i - \tilde{z}| = |x^i - z|$, and therefore $|\tilde{x}^j - \tilde{z}| = |x^j - z|$.

In the same way, the natural map can be defined for the spherical model triangle $\tilde{\triangle}(x^1x^2x^3)_{\mathbb{S}^2}$.

2.2.1. Definition of thin triangles. *A triangle $[x^1x^2x^3]$ in the metric space $\mathcal{X}$ is called* thin *if the natural map $\tilde{\triangle}(x^1x^2x^3)_{\mathbb{E}^2} \to [x^1x^2x^3]$ is short (that is, a distance nonincreasing map).*

Analogously, a triangle $[x^1x^2x^3]$ is called spherically thin *if the natural map from the spherical model triangle $\tilde{\triangle}(x^1x^2x^3)_{\mathbb{S}^2}$ to $[x^1x^2x^3]$ is short.*

2.2.2. Proposition. *A geodesic space is* CAT(0) *(*CAT(1)*) if and only if all its triangles are thin (respectively, all its triangles of perimeter $< 2\cdot\pi$ are spherically thin).*

Proof; "if" part. Apply the triangle inequality and thinness of triangles $[pxy]$ and $[qxy]$, where p, q, x, and y are as in the definition of CAT(κ) comparison (Section 2.1).

"Only if" part. Applying CAT(0) comparison to a quadruple p, q, x, y with $q \in [xy]$ shows that any triangle satisfies *point-side comparison*, that is, the distance from a vertex to a point on the opposite side is no greater than the corresponding distance in the Euclidean model triangle.

Now consider a triangle $[x^1x^2x^3]$, and let $y \in [x^1x^2]$ and $z \in [x^1x^3]$. Let $\tilde{y}, \tilde{z}$ be the corresponding points on the sides of the model triangle $\tilde{\triangle}(x^1x^2x^3)_{\mathbb{E}^2}$. Applying point-side comparison first to the triangle $[x^1x^2x^3]$ with $y \in [x^1x^2]$, and then to the triangle $[x^1yx^3]$ with $z \in [x^1x^3]$, implies that model angles satisfy

$$\tilde{\measuredangle}(x^1\,{}^{x^2}_{x^3})_{\mathbb{E}^2} \geqslant \tilde{\measuredangle}(x^1\,{}^{y}_{x^3})_{\mathbb{E}^2} \geqslant \tilde{\measuredangle}(x^1\,{}^{y}_{z})_{\mathbb{E}^2}.$$

Therefore $|\tilde{y} - \tilde{z}|_{\mathbb{E}^2} \geqslant |y - z|$.

The CAT(1) argument is the same. □

2.2.3. Uniqueness of geodesics. *In a proper length* CAT(0) *space, pairs of points are joined by unique geodesics, and these geodesics depend continuously on their endpoint pairs.*

Analogously, in a proper length CAT(1) *space, pairs of points at distance less than π are joined by unique geodesics, and these geodesics depend continuously on their endpoint pairs.*

Proof. Given 4 points p^1, p^2, q^1, q^2 in a proper length CAT(0) space $\mathcal{U}$, consider two triangles $[p^1q^1p^2]$ and $[p^2q^2q^1]$. Since both of these triangles are thin, we get

$$|\mathrm{path}_{[p^1q^1]}(t) - \mathrm{path}_{[p^2q^1]}(t)|_{\mathcal{U}} \leqslant (1-t)\cdot|p^1 - p^2|_{\mathcal{U}},$$
$$|\mathrm{path}_{[p^2q^1]}(t) - \mathrm{path}_{[p^2q^2]}(t)|_{\mathcal{U}} \leqslant t\cdot|q^1 - q^2|_{\mathcal{U}}.$$

By the triangle inequality,

$$|\mathrm{path}_{[p^1q^1]}(t) - \mathrm{path}_{[p^2q^2]}(t)|_{\mathcal{U}} \leqslant \max\{|p^1 - p^2|_{\mathcal{U}}, |q^1 - q^2|_{\mathcal{U}}\}.$$

Hence, continuity and uniqueness in the CAT(0) case follow.

The CAT(1) case is done in essentially the same way. □

Adding the first two inequalities of the preceding proof gives the following:

2.2.4. Proposition. *Suppose p^1, p^2, q^1, q^2 are points in a proper length* CAT(0) *space $\mathcal{U}$. Then*

$$|\mathrm{path}_{[p^1q^1]}(t) - \mathrm{path}_{[p^2q^2]}(t)|_{\mathcal{U}}$$

is a convex function.

2.2.5. Corollary. *Let K be a closed convex subset in a proper length* CAT(0) *space $\mathcal{U}$. Then* $\mathrm{dist}_K : \mathcal{U} \to \mathbb{R}$ is convex*; that is, the function $t \mapsto$* $\mathrm{dist}_K \circ \gamma$ *is convex for any geodesic γ in $\mathcal{U}$.*

In particular, dist_p *is convex for any point p in $\mathcal{U}$.*

2.2.6. Corollary. *Any proper length* CAT(0) *space is contractible.*

Analogously, any proper length CAT(1) *space with diameter $< \pi$ is contractible.*

Proof. Let $\mathcal{U}$ be a proper length CAT(0) space. Fix a point $p \in \mathcal{U}$.

For each point x consider the geodesic path $\gamma_x \colon [0, 1] \to \mathcal{U}$ from p to x. Consider the one-parameter family of maps $h_t \colon x \mapsto \gamma_x(t)$ for $t \in [0, 1]$. By uniqueness of geodesics (2.2.3), the map $(t, x) \mapsto h_t(x)$ is continuous. The family h_t is called a *geodesic homotopy*.

It remains to note that $h_1(x) = x$ and $h_0(x) = p$ for any x.

The proof of the CAT(1) case is identical. □

2.2.7. Proposition. *Suppose $\mathcal{U}$ is a proper length* CAT(0) *space. Then any local geodesic in $\mathcal{U}$ is a geodesic.*

Analogously, if $\mathcal{U}$ is a proper length CAT(1) *space, then any local geodesic in $\mathcal{U}$ which is shorter than π is a geodesic.*

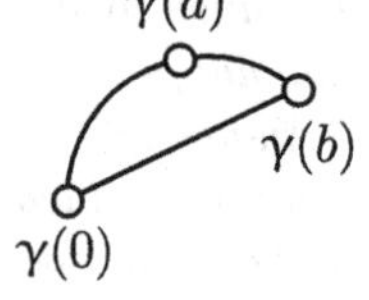

Proof. Suppose $\gamma \colon [0, \ell] \to \mathcal{U}$ is a local geodesic that is not a geodesic. Choose a to be the maximal value such that γ is a geodesic on $[0, a]$. Further choose $b > a$ so that γ is a geodesic on $[a, b]$.

Since the triangle $[\gamma(0)\gamma(a)\gamma(b)]$ is thin and $|\gamma(0) - \gamma(b)| < b$, we have

$$|\gamma(a - \varepsilon) - \gamma(a + \varepsilon)| < 2 \cdot \varepsilon$$

for all small $\varepsilon > 0$. That is, γ is not length-minimizing on the interval $[a - \varepsilon, a + \varepsilon]$ for any $\varepsilon > 0$, a contradiction.

The spherical case is done in the same way. □

2.2.8. Exercise. *Assume $\mathcal{U}$ is a proper length* CAT(κ) *space with extendable geodesics; that is, any geodesic is an arc in a local geodesic $\mathbb{R} \to \mathcal{U}$.*

Show that the space of geodesic directions at any point in $\mathcal{U}$ is complete.

Does the statement remain true if $\mathcal{U}$ is complete, but not required to be proper?

Now let us formulate the main result of this section.

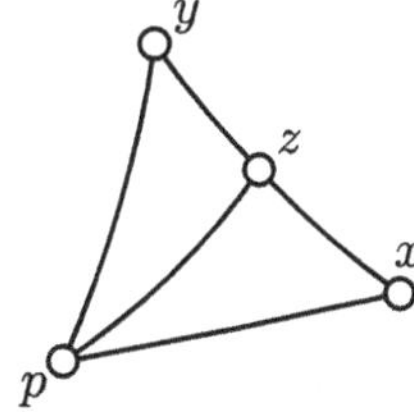

2.2.9. Inheritance lemma. *Assume that a triangle* $[pxy]$ *in a metric space is* decomposed *into two triangles* $[pxz]$ *and* $[pyz]$*; that is,* $[pxz]$ *and* $[pyz]$ *have a common side* $[pz]$*, and the sides* $[xz]$ *and* $[zy]$ *together form the side* $[xy]$ *of* $[pxy]$*.*

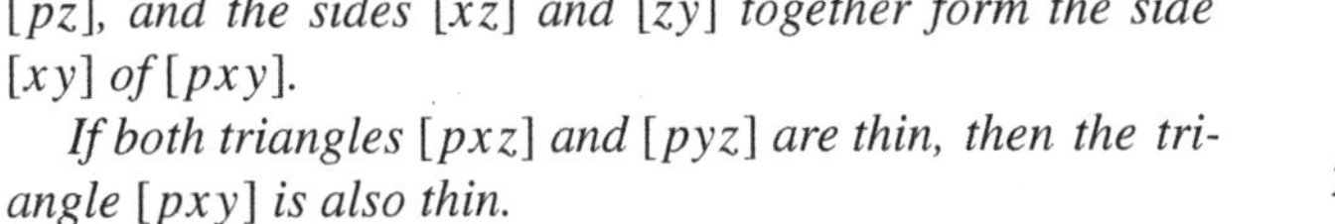

If both triangles $[pxz]$ *and* $[pyz]$ *are thin, then the triangle* $[pxy]$ *is also thin.*

Analogously, if $[pxy]$ *has perimeter* $< 2\cdot\pi$ *and both triangles* $[pxz]$ *and* $[pyz]$ *are spherically thin, then triangle* $[pxy]$ *is spherically thin.*

Proof. Construct the model triangles $[\dot p\dot x\dot z] = \tilde\triangle(pxz)_{\mathbb{E}^2}$ and $[\dot p\dot y\dot z] = \tilde\triangle(pyz)_{\mathbb{E}^2}$ so that $\dot x$ and $\dot y$ lie on opposite sides of $[\dot p\dot z]$.

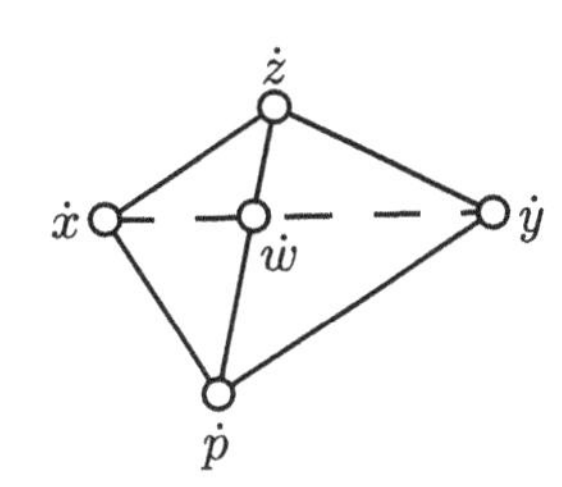

Let us show that

❶ $$\tilde\measuredangle(z\,{}^p_x) + \tilde\measuredangle(z\,{}^p_y) \geqslant \pi.$$

Suppose the contrary, that is

$$\tilde\measuredangle(z\,{}^p_x) + \tilde\measuredangle(z\,{}^p_y) < \pi.$$

Then for some point $\dot w \in [\dot p\dot z]$, we have

$$|\dot x - \dot w| + |\dot w - \dot y| < |\dot x - \dot z| + |\dot z - \dot y| = |x - y|.$$

Let $w \in [pz]$ correspond to $\dot w$; that is, $|z - w| = |\dot z - \dot w|$. Since $[pxz]$ and $[pyz]$ are thin, we have

$$|x - w| + |w - y| < |x - y|,$$

contradicting the triangle inequality.

Denote by $\dot D$ the union of two solid triangles $[\dot p\dot x\dot z]$ and $[\dot p\dot y\dot z]$. Further, denote by $\tilde D$ the solid triangle $[\tilde p\tilde x\tilde y] = \tilde\triangle(pxy)_{\mathbb{E}^2}$. By ❶, there is a short map $F\colon \tilde D \to \dot D$ that sends

$$\tilde p \mapsto \dot p, \qquad \tilde x \mapsto \dot x, \qquad \tilde z \mapsto \dot z, \qquad \tilde y \mapsto \dot y. \qquad \square$$

2.2.10. Exercise. *Use Alexandrov's lemma (1.5.1) to prove the last statement.*

By assumption, the natural maps $[\dot p\dot x\dot z] \to [pxz]$ and $[\dot p\dot y\dot z] \to [pyz]$ are short. By composition, the natural map from $[\tilde p\tilde x\tilde y]$ to $[pyz]$ is short, as claimed.

The spherical case is done along the same lines. $\square$

2.2.11. Exercise. *Show that any ball in a proper length* CAT(0) *space is a convex set.*

Analogously, show that any ball of radius $R < \frac{\pi}{2}$ in a proper length CAT(1) *space is a convex set.*

Recall that a set A in a metric space $\mathcal{U}$ is called locally convex if for any point $p \in A$ there is an open neighborhood $\mathcal{U} \ni p$ such that any geodesic in $\mathcal{U}$ with ends in A lies in A.

2.2.12. Exercise. *Let $\mathcal{U}$ be a proper length* CAT(0) *space. Show that any closed, connected, locally convex set in $\mathcal{U}$ is convex.*

2.2.13. Exercise. *Let $\mathcal{U}$ be a proper length* CAT(0) *space and $K \subset \mathcal{U}$ be a closed convex set. Show that:*

(a) *For each point $p \in \mathcal{U}$ there is unique point $p^* \in K$ that minimizes the distance $|p - p^*|$.*
(b) *The closest-point projection $p \mapsto p^*$ defined by (a) is short.*

2.3 Reshetnyak's gluing theorem

Suppose $\mathcal{U}^1$ and $\mathcal{U}^2$ are proper length spaces with isometric closed convex sets $A^i \subset \mathcal{U}^i$, and let $\iota \colon A^1 \to A^2$ be an isometry. Consider the space $\mathcal{W}$ of all equivalence classes in $\mathcal{U}^1 \sqcup \mathcal{U}^2$ with the equivalence relation given by $a \sim \iota(a)$ for any $a \in A^1$.

It is straightforward to see that $\mathcal{W}$ is a proper length space when equipped with the following metric

$$\begin{aligned} |x - y|_{\mathcal{W}} &\stackrel{def}{=\!=} |x - y|_{\mathcal{U}^i} \\ &\quad \text{if} \;\; x, y \in \mathcal{U}^i, \quad \text{and} \\ |x - y|_{\mathcal{W}} &\stackrel{def}{=\!=} \min \left\{ |x - a|_{\mathcal{U}^1} + |y - \iota(a)|_{\mathcal{U}^2} : a \in A^1 \right\} \\ &\quad \text{if} \;\; x \in \mathcal{U}^1 \;\; \text{and} \;\; y \in \mathcal{U}^2. \end{aligned}$$

Abusing notation, we denote by x and y the points in $\mathcal{U}^1 \sqcup \mathcal{U}^2$ and their equivalence classes in $\mathcal{U}^1 \sqcup \mathcal{U}^2/\sim$.

The space $\mathcal{W}$ is called the *gluing* of $\mathcal{U}^1$ and $\mathcal{U}^2$ along ι. If one applies this construction to two copies of one space $\mathcal{U}$ with a set $A \subset \mathcal{U}$ and the identity map $\iota \colon A \to A$, then the obtained space is called the *double* of $\mathcal{U}$ along A.

We can (and will) identify $\mathcal{U}^i$ with its image in $\mathcal{W}$; this way both subsets $A^i \subset \mathcal{U}^i$ will be identified and denoted further by A. Note that $A = \mathcal{U}^1 \cap \mathcal{U}^2 \subset \mathcal{W}$, therefore A is also a convex set in $\mathcal{W}$.

The following theorem was proved by Yuri Reshetnyak in [57].

2.3.1. Reshetnyak gluing. *Suppose $\mathcal{U}^1$ and $\mathcal{U}^2$ are proper length* CAT(0) *spaces with isometric closed convex sets $A^i \subset \mathcal{U}^i$, and $\iota\colon A^1 \to A^2$ is an isometry. Then the gluing of $\mathcal{U}^1$ and $\mathcal{U}^2$ along ι is a* CAT(0) *proper length space.*

Proof. By construction of the gluing space, the statement can be reformulated in the following way.

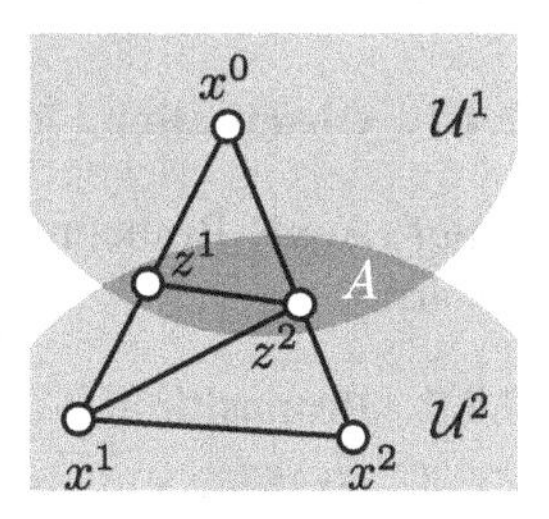

2.3.2. Reformulation of 2.3.1. *Let $\mathcal{W}$ be a proper length space which has two closed convex sets $\mathcal{U}^1, \mathcal{U}^2 \subset \mathcal{W}$ such that $\mathcal{U}^1 \cup \mathcal{U}^2 = \mathcal{W}$ and $\mathcal{U}^1, \mathcal{U}^2$ are* CAT(0). *Then $\mathcal{W}$ is* CAT(0).

It suffices to show that any triangle $[x^0x^1x^2]$ in $\mathcal{W}$ is thin. This is obviously true if all three points x^0, x^1, x^2 lie in one of $\mathcal{U}^i$. Thus, without loss of generality, we may assume that $x^0 \in \mathcal{U}^1$ and $x^1, x^2 \in \mathcal{U}^2$.

Choose points $z^1, z^2 \in A = \mathcal{U}^1 \cap \mathcal{U}^2$ that lie respectively on the sides $[x^0x^1]$, $[x^0x^2]$. Note that

- the triangle $[x^0z^1z^2]$ lies in $\mathcal{U}^1$,
- both triangles $[x^1z^1z^2]$ and $[x^1z^2x^2]$ lie in $\mathcal{U}^2$.

In particular each triangle $[x^0z^1z^2]$, $[x^1z^1z^2]$ and $[x^1z^2x^2]$ is thin.

Applying the inheritance lemma for thin triangles (2.2.9) twice, we get that $[x^0x^1z^2]$ and consequently $[x^0x^1x^2]$ is thin. □

2.4 Reshetnyak puff pastry

In this section we introduce the notion of Reshetnyak puff pastry. This construction will be used in the next section to prove the Collision theorem (2.6.1).

Let $\boldsymbol{A} = (A^1, \dots, A^N)$ be an array of convex closed sets in the Euclidean space $\mathbb{E}^m$. Consider an array of $N+1$ copies of $\mathbb{E}^m$. Assume that the space $\mathcal{R}$ is obtained by gluing successive pairs of spaces along $A^1, \dots, A^N$, respectively.

The resulting space $\mathcal{R}$ will be called the *Reshetnyak puff pastry* for the array $\boldsymbol{A}$. The copies of $\mathbb{E}^m$ in the puff pastry $\mathcal{R}$ will be called *levels*; they will be denoted by $\mathcal{R}^0, \dots, \mathcal{R}^N$. The point in the kth level $\mathcal{R}^k$ that corresponds to $x \in \mathbb{E}^m$ will be denoted by x^k.

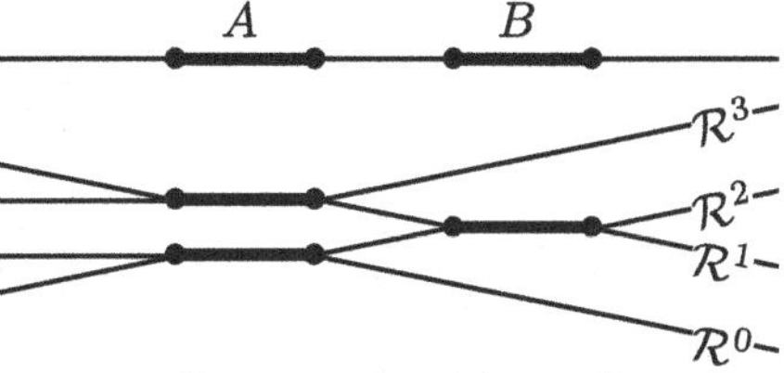

Puff pastry for (A, B, A).

Given $x \in \mathbb{E}^m$, any point $x^k \in \mathcal{R}$ is called a *lifting* of x. The map $x \mapsto x^k$ defines an isometry $\mathbb{E}^m \to \mathcal{R}^k$; in particular we can talk about liftings of subsets in $\mathbb{E}^m$.

Note that:

- The intersection $A^1 \cap \dots \cap A^N$ admits a unique lifting in $\mathcal{R}$.
- Moreover, $x^i = x^j$ for some $i < j$ if and only if

$$x \in A^{i+1} \cap \dots \cap A^j.$$

- The restriction $\mathcal{R}^k \to \mathbb{E}^m$ of the natural projection $x^k \mapsto x$ is an isometry.

2.4.1. Observation. *Any Reshetnyak puff pastry is a proper length* CAT(0) *space.*

Proof. Apply Reshetnyak gluing theorem (2.3.1) recursively for the convex sets in the array. □

2.4.2. Proposition. *Assume $(A^1, \dots, A^N)$ and $(\check{A}^1, \dots, \check{A}^N)$ are two arrays of convex closed sets in $\mathbb{E}^m$ such that $A^k \subset \check{A}^k$ for each k. Let $\mathcal{R}$ and $\check{\mathcal{R}}$ be the corresponding Reshetnyak puff pastries. Then the map $\mathcal{R} \to \check{\mathcal{R}}$ defined by $x^k \mapsto \check{x}^k$ is short.*

Moreover, if

$$|x^i - y^j|_{\mathcal{R}} = |\check{x}^i - \check{y}^j|_{\check{\mathcal{R}}} \qquad \text{❶}$$

for some $x, y \in \mathbb{E}^m$ and $i, j \in \{0, \dots, n\}$, then the unique geodesic $[\check{x}^i \check{y}^j]_{\check{\mathcal{R}}}$ is the image of the unique geodesic $[x^i y^j]_{\mathcal{R}}$ under the map $x^i \mapsto \check{x}^i$.

Proof. The first statement in the proposition follows from the construction of Reshetnyak puff pastries.

By Observation 2.4.1, $\mathcal{R}$ and $\check{\mathcal{R}}$ are proper length CAT(0) spaces; hence, $[x^i y^j]_{\mathcal{R}}$ and $[\check{x}^i \check{y}^j]_{\check{\mathcal{R}}}$ are unique. By ❶, since the map $\mathcal{R} \to \check{\mathcal{R}}$ is short, the image of $[x^i y^j]_{\mathcal{R}}$ is a geodesic of $\check{\mathcal{R}}$ joining $\check{x}^i$ to $\check{y}^j$. Hence the second statement follows. □

2.4.3. Definition. *Consider a Reshetnyak puff pastry $\mathcal{R}$ with the levels $\mathcal{R}^0, \dots, \mathcal{R}^N$. We say that $\mathcal{R}$ is* end-to-end convex *if $\mathcal{R}^0 \cup \mathcal{R}^N$, the union of its lower and upper levels, forms a convex set in $\mathcal{R}$.*

Note that if $\mathcal{R}$ is the Reshetnyak puff pastry for an array of convex sets $\boldsymbol{A} = (A^1, \dots, A^N)$, then $\mathcal{R}$ is end-to-end convex if and only if the union of the lower and the upper levels $\mathcal{R}^0 \cup \mathcal{R}^N$ is isometric to the double of $\mathbb{E}^m$ along the nonempty intersection $A^1 \cap \dots \cap A^N$.

2.4.4. Observation. *Let $\check{\boldsymbol{A}}$ and $\boldsymbol{A}$ be arrays of convex bodies in $\mathbb{E}^m$. Assume that the array $\boldsymbol{A}$ is obtained by inserting in $\check{\boldsymbol{A}}$ several copies of the bodies which were already listed in $\check{\boldsymbol{A}}$.*

For example, if $\check{\boldsymbol{A}} = (A, C, B, C, A)$, by placing B in the second place and A in the fourth place, we obtain $\boldsymbol{A} = (A, B, C, A, B, C, A)$.

Denote by $\check{\mathcal{R}}$ and $\mathcal{R}$ the Reshetnyak puff pastries for $\check{\boldsymbol{A}}$ and $\boldsymbol{A}$, respectively.

If $\check{\mathcal{R}}$ is end-to-end convex, then so is $\mathcal{R}$.

Proof. Without loss of generality we may assume that $\boldsymbol{A}$ is obtained by inserting one element in $\check{\boldsymbol{A}}$, say at the place number k.

Note that $\check{\mathcal{R}}$ is isometric to the puff pastry for $\boldsymbol{A}$ with A^k replaced by $\mathbb{E}^m$. It remains to apply Proposition 2.4.2. □

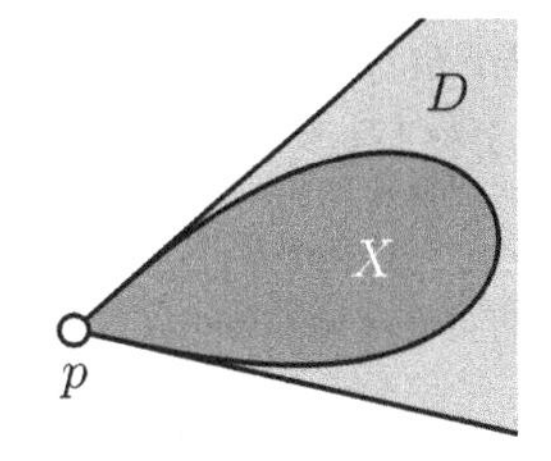

Let X be a convex set in a Euclidean space. By a *dihedral angle*, we understand an intersection of two half-spaces; the intersection of corresponding hyperplanes is called the *edge* of the angle. We say that a dihedral angle D supports X at a point $p \in X$ if D contains X and the edge of D contains p.

2.4.5. Lemma. *Let A and B be two convex sets in $\mathbb{E}^m$. Assume that any dihedral angle supporting $A \cap B$ at some point has angle measure at least α. Then the Reshetnyak puff pastry for the array*

$$(\underbrace{A, B, A, \dots}_{\lceil\frac{\pi}{\alpha}\rceil + 1 \text{ times}}).$$

is end-to-end convex.

The proof of the lemma is based on a partial case, which we formulate as a sublemma.

2.4.6. Sublemma. *Let $\ddot{A}$ and $\ddot{B}$ be two half-planes in $\mathbb{E}^2$, where $\ddot{A} \cap \ddot{B}$ is an angle with measure α. Then the Reshetnyak puff pastry for the array*

$$(\underbrace{\ddot{A}, \ddot{B}, \ddot{A}, \dots}_{\lceil\frac{\pi}{\alpha}\rceil + 1 \text{ times}})$$

is end-to-end convex.

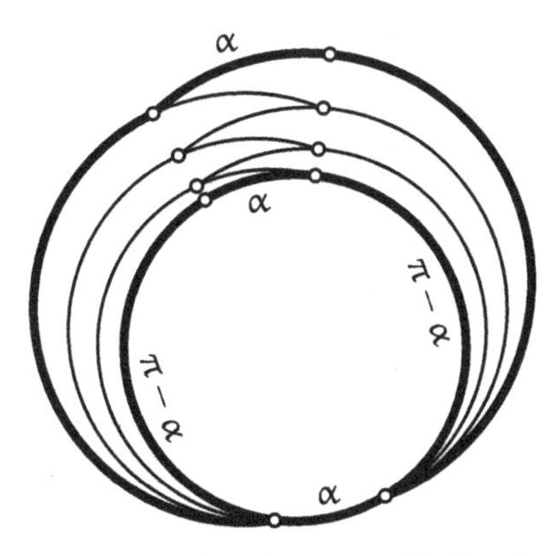

Proof. Note that the puff pastry $\ddot{\mathcal{R}}$ is isometric to the cone over the space glued from the unit circles as shown on the diagram.

All the short arcs on the diagram have length α; the long arcs have length $\pi - \alpha$, so making a circuit along any path will take $2 \cdot \pi$.

Observe that end-to-end convexity of $\ddot{\mathcal{R}}$ is equivalent to the fact that any geodesic shorter than π with the ends on the inner and the outer circles lies completely in the union of these two circles.

The latter holds if the zigzag line in the picture has length at least π. This line is formed by $\lceil\frac{\pi}{\alpha}\rceil$ arcs with length α each. Hence the sublemma. □

In the proof of 2.4.5, we will use the following exercise in convex geometry:

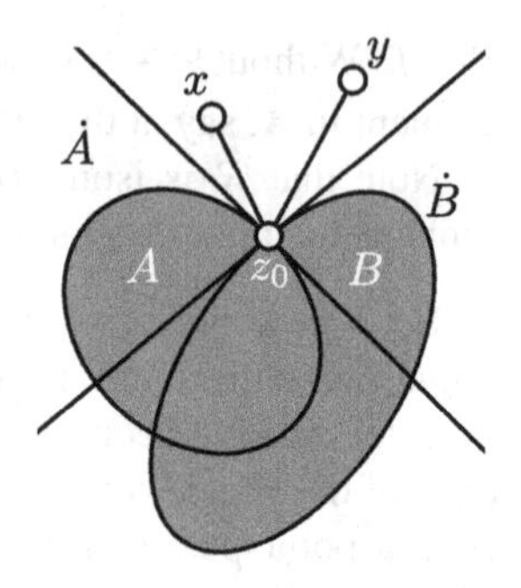

2.4.7. Exercise. *Let A and B be two closed convex sets in $\mathbb{E}^m$ and $A \cap B \neq \emptyset$. Given two points $x, y \in \mathbb{E}^m$ let $f(z) = |x - z| + |y - z|$.*

Let $z_0 \in A \cap B$ be a point of minimum of $f|_{A\cap B}$.

Show that there are half-spaces $\dot{A}$ and $\dot{B}$ such that $\dot{A} \supset A$ and $\dot{B} \supset B$, and z_0 is also a point of minimum of the restriction $f|_{\dot{A}\cap\dot{B}}$.

Proof of 2.4.5. Fix arbitrary $x, y \in \mathbb{E}^m$. Choose a point $z \in A \cap B$ for which the sum

$$|x - z| + |y - z|$$

is minimal. To show the end-to-end convexity of $\mathcal{R}$, it is sufficient to prove the following:

❷ *The geodesic* $[x^0 y^N]_{\mathcal{R}}$ *contains* $z^0 = z^N \in \mathcal{R}$.

Without loss of generality we may assume that $z \in \partial A \cap \partial B$. Indeed, since the puff pastry for the 1-array (B) is end-to-end convex, Proposition 2.4.2 together with Observation 2.4.4 imply ❷ in case z lies in the interior of A. In the same way we can treat the case when z lies in the interior of B.

Note that $\mathbb{E}^m$ admits an isometric splitting $\mathbb{E}^{m-2} \times \mathbb{E}^2$ such that

$$\dot{A} = \mathbb{E}^{m-2} \times \ddot{A}$$
$$\dot{B} = \mathbb{E}^{m-2} \times \ddot{B}$$

where $\ddot{A}$ and $\ddot{B}$ are half-planes in $\mathbb{E}^2$.

Using Exercise 2.4.7, let us replace each A by $\dot{A}$ and each B by $\dot{B}$ in the array, to get the array

$$(\underbrace{\dot{A}, \dot{B}, \dot{A}, \dots}_{\lceil\frac{\pi}{\alpha}\rceil + 1 \text{ times}}).$$

The corresponding puff pastry $\dot{\mathcal{R}}$ splits as a product of $\mathbb{E}^{m-2}$ and a puff pastry, call it $\ddot{\mathcal{R}}$, glued from the copies of the plane $\mathbb{E}^2$ for the array

$$(\underbrace{\ddot{A}, \ddot{B}, \ddot{A}, \dots}_{\lceil\frac{\pi}{\alpha}\rceil + 1 \text{ times}}).$$

Note that the dihedral angle $\dot{A} \cap \dot{B}$ is at least α. Therefore the angle measure of $\ddot{A} \cap \ddot{B}$ is also at least α. According to Sublemma 2.4.6 and Observation 2.4.4, $\ddot{\mathcal{R}}$ is end-to-end convex.

Since $\dot{\mathcal{R}} \overset{iso}{=\!=} \mathbb{E}^{m-2} \times \ddot{\mathcal{R}}$, the puff pastry $\dot{\mathcal{R}}$ is also end-to-end convex.

It follows that the geodesic $[\dot{x}^0 \dot{y}^N]_{\dot{\mathcal{R}}}$ contains $\dot{z}^0 = \dot{z}^N \in \dot{\mathcal{R}}$. By Proposition 2.4.2, the image of $[\dot{x}^0 \dot{y}^N]_{\dot{\mathcal{R}}}$ under the map $\dot{x}^k \mapsto x^k$ is the geodesic $[x^0 y^N]_{\mathcal{R}}$. Hence Claim ❷ and the lemma follow. □

2.5 Wide corners

We say that a closed convex set $A \subset \mathbb{E}^m$ has *ε-wide corners* for given $\epsilon > 0$ if together with each point p, the set A contains a small right circular cone with tip at p and aperture ε; that is, ε is the maximum angle between two generating lines of the cone.

For example, a plane polygon has ε-wide corners if all its interior angles are at least ε.

We will consider finite collections of closed convex sets $A^1, \dots, A^n \subset \mathbb{E}^m$ such that for any subset $F \subset \{1, \dots, n\}$, the intersection $\bigcap_{i \in F} A^i$ has ε-wide corners. In this case we may say briefly *all intersections of A^i have ε-wide corners.*

2.5.1. Exercise. *Assume $A^1, \dots, A^n \subset \mathbb{E}^m$ are compact, convex sets with a common interior point. Show that all intersections of A^i have ε-wide corners for some positive ε.*

2.5.2. Exercise. *Assume $A^1, \dots, A^n \subset \mathbb{E}^m$ are convex sets with nonempty interior that have a common center of symmetry. Show that all intersections of A^i have ε-wide corners for some positive ε.*

The proof of the following proposition is based on Lemma 2.4.5; this lemma is essentially the case $n = 2$ in the proposition.

2.5.3. Proposition. *Given $\varepsilon > 0$ and a positive integer n, there is an array of integers $\boldsymbol{j}_\varepsilon(n) = (j_1, \dots, j_N)$ such that:*

(a) For each k we have $1 \leqslant j_k \leqslant n$, and each number $1, \dots, n$ appears in $\boldsymbol{j}_\varepsilon$ at least once.

(b) If $A^1, \dots, A^n$ is a collection of closed convex sets in $\mathbb{E}^m$ with a common point and all their intersections have ε-wide corners, then the puff pastry for the array $(A^{j_1}, \dots, A^{j_N})$ is end-to-end convex.

Moreover we can assume that $N \leqslant (\lceil \frac{\pi}{\varepsilon} \rceil + 1)^n$.

Proof. The array $\boldsymbol{j}_\varepsilon(n) = (j_1, \dots, j_N)$ is constructed recursively. For $n = 1$, we can take $\boldsymbol{j}_\varepsilon(1) = (1)$.

Assume that $\boldsymbol{j}_\varepsilon(n)$ is constructed. Let us replace each occurrence of n in $\boldsymbol{j}_\varepsilon(n)$ by the alternating string

$$\underbrace{n, n+1, n, \dots}_{\lceil \frac{\pi}{\varepsilon} \rceil + 1 \text{ times}}.$$

Denote the obtained array by $\boldsymbol{j}_\varepsilon(n+1)$.

By Lemma 2.4.5, end-to-end convexity of the puff pastry for $\boldsymbol{j}_\varepsilon(n+1)$ follows from end-to-end convexity of the puff pastry for the array where each string

$$\underbrace{A^n, A^{n+1}, A^n, \dots}_{\left\lceil\frac{\pi}{\varepsilon}\right\rceil+1 \text{ times}}$$

is replaced by $Q = A^n \cap A^{n+1}$. End-to-end convexity of the latter follows by the assumption on $\boldsymbol{j}_\varepsilon(n)$, since all the intersections of $A^1, \dots, A^{n-1}, Q$ have ε-wide corners.

The upper bound on N follows directly from the construction. □

2.6 Billiards

Let $A^1, A^2, \dots A^n$ be a finite collection of closed convex sets in $\mathbb{E}^m$. Assume that for each i the boundary ∂A^i is a smooth hypersurface.

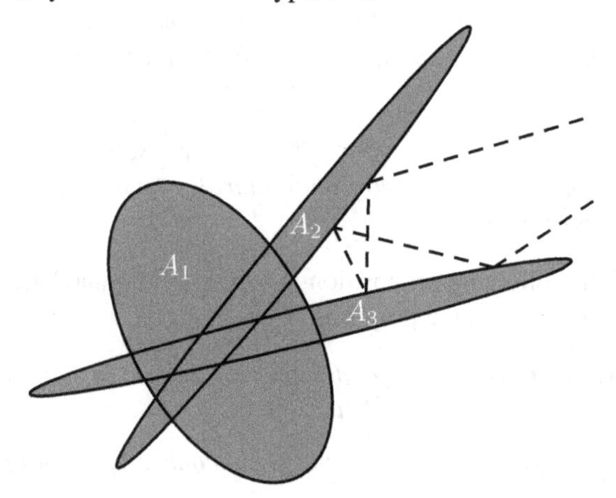

Consider the billiard table formed by the closure of the complement

$$T = \overline{\mathbb{E}^m \backslash \bigcup_i A^i}.$$

The sets A^i will be called *walls* of the table T, and the billiards described above will be called *billiards with convex walls*.

A *billiard trajectory* on the table T is a unit speed broken line γ that follows the standard law of billiards at the break points on ∂A^i—in particular, the angle of reflection is equal to the angle of incidence. The break points of the trajectory will be called *collisions*. We assume the trajectory meets only one wall at a time.

Recall that the definition of sets with ε-wide corners is given in Section 2.5.

2.6.1. Collision theorem. *Assume $T \subset \mathbb{E}^m$ is a billiard table with n convex walls. Assume that the walls of T have a common interior point and all their intersections have ε-wide corners. Then the number of collisions of any trajectory in T is bounded by a number N which depends only on n and ε.*

As we will see from the proof, the value N can be found explicitly;

$$N = (\lceil \tfrac{\pi}{\varepsilon} \rceil + 1)^{n^2}$$

will do.

The collision theorem was proved by Dmitri Burago, Serge Ferleger, and Alexey Kononenko in [21]; we present their proof with minor improvements.

Let us formulate and prove a corollary of the collision theorem; it answers a question formulated by Yakov Sinai [37].

2.6.2. Corollary. *Consider n homogeneous hard balls moving freely and colliding elastically in $\mathbb{R}^3$. Every ball moves along a straight line with constant speed until two balls collide, and then the new velocities of the two balls are determined by the laws of classical mechanics. We assume that only two balls can collide at the same time.*

Then the total number of collisions cannot exceed some number N that depends on the radii and masses of the balls. If the balls are identical, then N depends only on n.

The proof below admits a straightforward generalization to all dimensions.

Proof. Denote by $a_i = (x_i, y_i, z_i) \in \mathbb{R}^3$ the center of the i-th ball. Consider the corresponding point in $\mathbb{R}^{3 \cdot N}$

$$\begin{aligned} \boldsymbol{a} &= (a_1, a_2, \dots, a_n) = \\ &= (x_1, y_1, z_1, x_2, y_2, z_2, \dots, x_n, y_n, z_n). \end{aligned}$$

The i-th and j-th ball intersect if

$$|a_i - a_j| \leqslant R_i + R_j,$$

where R_i denotes the radius of the i-th ball. These inequalities define $\frac{n \cdot (n-1)}{2}$ cylinders

$$C_{i,j} = \left\{ (a_1, a_2, \dots, a_n) \in \mathbb{R}^{3 \cdot n} : |a_i - a_j| \leqslant R_i + R_j \right\}.$$

The closure of the complement

$$T = \overline{\mathbb{R}^{3 \cdot n} \backslash \bigcup_{i<j} C_{i,j}}$$

is the configuration space of our system. Its points correspond to valid positions of the system of balls.

The evolution of the system of balls is described by the motion of the point $\boldsymbol{a} \in \mathbb{R}^{3 \cdot n}$. It moves along a straight line at a constant speed until it hits one of the cylinders $C_{i,j}$; this event corresponds to a collision in the system of balls.

Consider the norm of $\boldsymbol{a} = (a_1, \dots, a_n) \in \mathbb{R}^{3 \cdot n}$ defined by

$$\|\boldsymbol{a}\| = \sqrt{M_1 \cdot |a_1|^2 + \dots + M_n \cdot |a_n|^2},$$

where $|a_i| = \sqrt{x_i^2 + y_i^2 + z_i^2}$ and M_i denotes the mass of the i-th ball. In the metric defined by $\|*\|$, the collisions follow the standard law of billiards.

By construction, the number of collisions of hard balls that we need to estimate is the same as the number of collisions of the corresponding billiard trajectory on the table T with $C_{i,j}$ as the walls.

Note that each cylinder $C_{i,j}$ is a convex set; it has smooth boundary, and it is centrally symmetric around the origin. By Exercise 2.5.2, all the intersections of the walls have ε-wide corners for some $\varepsilon > 0$ that depend on the radiuses R_i and the masses M_i. It remains to apply the collision theorem (2.6.1). □

Now we present the proof of the collision theorem (2.6.1) based on the results developed in the previous section.

Proof of 2.6.1. Let us apply induction on n.

Base: $n = 1$. The number of collisions cannot exceed 1. Indeed, by the convexity of A^1, if the trajectory is reflected once in ∂A^1, then it cannot return to A^1.

Step. Assume γ is a trajectory that meets the walls in the order $A^{i_1}, \dots, A^{i_N}$ for a large integer N.

Consider the array

$$\boldsymbol{A}_\gamma = (A^{i_1}, \dots, A^{i_N}).$$

The induction hypothesis implies:

❶ *There is a positive integer M such that any M consecutive elements of $\boldsymbol{A}_\gamma$ contain each A^i at least once.*

Let $\mathcal{R}_\gamma$ be the Reshetnyak puff pastry for $\boldsymbol{A}_\gamma$.

Consider the lift of γ to $\mathcal{R}_\gamma$, defined by $\bar{\gamma}(t) = \gamma^k(t) \in \mathcal{R}_\gamma$ for any moment of time t between the kth and $(k+1)$th collisions. Since γ follows the standard law of billiards at break points, the lift $\bar{\gamma}$ is locally a geodesic in $\mathcal{R}_\gamma$. By Observation 2.4.1, the puff pastry $\mathcal{R}_\gamma$ is a proper length CAT(0) space. Therefore $\bar{\gamma}$ is a geodesic.

Since γ does not meet $A^1 \cap \cdots \cap A^n$, the lift $\bar{\gamma}$ does not lie in $\mathcal{R}^0_\gamma \cup \mathcal{R}^N_\gamma$. In particular, $\mathcal{R}_\gamma$ is not end-to-end convex.

Let

$$\boldsymbol{B} = (A^{j_1}, \ldots, A^{j_K})$$

be the array provided by Proposition 2.5.3; so $\boldsymbol{B}$ contains each A^i at least once and the puff pastry $\mathcal{R}_{\boldsymbol{B}}$ for $\boldsymbol{B}$ is end-to-end convex. If N is sufficiently large, namely $N \geqslant K \cdot M$, then ❶ implies that $\boldsymbol{A}_\gamma$ can be obtained by inserting a finite number of A^i's in $\boldsymbol{B}$.

By Observation 2.4.4, $\mathcal{R}_\gamma$ is end-to-end convex, a contradiction. □

2.7 Comments

The gluing theorem (2.3.1) extends to the class of geodesic CAT(0) spaces, which by Exercise 2.1.4 includes all complete length CAT(0) spaces. It also admits a natural generalization to length CAT(κ) spaces; see the book of Martin Bridson and André Haefliger [18] and our book [6] for details.

Puff pastry is used to bound topological entropy of the billiard flow and to approximate the shortest billiard path that touches given lines in a given order; see the papers of Dmitri Burago with Serge Ferleger and Alexey Kononenko [22], and with Dimitri Grigoriev and Anatol Slissenko [23]. The lecture [19] gives a short survey on the subject.

Note that the interior points of the walls play a key role in the proof despite the fact that trajectories never go inside the walls. In a similar fashion, puff pastry was used by the first author and Richard Bishop in [4] to find the upper curvature bound for warped products.

In [41], Joel Hass constructed an example of a Riemannian metric on the 3-ball with negative curvature and concave boundary. This example might decrease your appetite for generalizing the collision theorem—while locally such a 3-ball looks as good as the billiards table in the theorem, the number of collisions is obviously infinite.

It was shown by Dmitri Burago and Sergei Ivanov [20] that the number of collisions that may occur between n identical balls $\mathbb{R}^3$ grows at least exponentially in n.

Chapter 3
Globalization and asphericity

In this chapter we introduce locally CAT(0) spaces and prove the globalization theorem that provides a sufficient condition for locally CAT(0) spaces to be globally CAT(0).

The theorem implies in particular that the universal metric cover of a proper length, locally CAT(0) space, is a proper length CAT(0) space. It follows that any proper length, locally CAT(0) space is aspherical; that is, its universal cover is contractible.

This globalization theorem leads to a *construction toy set*, described by the flag condition (3.5.5). Playing with this toy set, we produce examples of exotic aspherical spaces.

3.1 Locally CAT spaces

We say that a space $\mathcal{U}$ is *locally* CAT(0) (or *locally* CAT(1)) if a small closed ball centered at any point p in $\mathcal{U}$ is CAT(0) (or CAT(1), respectively).

For example, $\mathbb{S}^1$ is locally isometric to $\mathbb{R}$, and so $\mathbb{S}^1$ is locally CAT(0). On the other hand, $\mathbb{S}^1$ is not CAT(0), since closed local geodesics in $\mathbb{S}^1$ are not geodesics, so $\mathbb{S}^1$ does not satisfy Proposition 2.2.7.

If $\mathcal{U}$ is a proper length space, then it is locally CAT(0) (or locally CAT(1)) if and only if each point $p \in \mathcal{U}$ admits an open neighborhood Ω that is geodesic and such that any triangle in Ω is thin (or spherically thin, respectively). The proof goes along the same lines as in Exercise 2.2.11.

S. Alexander et al., *An Invitation to Alexandrov Geometry*,
SpringerBriefs in Mathematics, https://doi.org/10.1007/978-3-030-05312-3_3

3.2 Space of local geodesic paths

In this section we will study behavior of local geodesics in locally CAT(κ) spaces. The results will be used in the proof of the globalization theorem (3.3.1).

Recall that a *path* is a curve parametrized by [0, 1]. The space of paths in a metric space $\mathcal{U}$ comes with the natural metric

➊ $$|\alpha - \beta| = \sup\{\, |\alpha(t) - \beta(t)|_{\mathcal{U}} : t \in [0,1]\,\}.$$

3.2.1. Proposition. *Let $\mathcal{U}$ be a proper, locally* CAT(κ) *length space.*

Assume $\gamma_n\colon [0,1] \to \mathcal{U}$ is a sequence of local geodesic paths converging to a path $\gamma_\infty\colon [0,1] \to \mathcal{U}$. Then γ_∞ is a local geodesic path. Moreover

$$\operatorname{length}\gamma_n \to \operatorname{length}\gamma_\infty$$

as $n \to \infty$.

Proof. CAT(0) case. Fix $t \in [0,1]$. Let $R > 0$ be sufficiently small, so that $\overline{\mathrm{B}}[\gamma_\infty(t), R]$ forms a proper length CAT(0) space.

Assume that a local geodesic σ is shorter than $R/2$ and intersects the ball $\mathrm{B}(\gamma_\infty(t), R/2)$. Then σ cannot leave the ball $\overline{\mathrm{B}}[\gamma_\infty(t), R]$. Hence, by Proposition 2.2.7, σ is a geodesic. In particular, for all sufficiently large n, any arc of γ_n of length $R/2$ or less containing $\gamma_n(t)$ is a geodesic.

Since $\mathcal{B} = \overline{\mathrm{B}}[\gamma_\infty(t), R]$ is a proper length CAT(0) space, by Theorem 2.2.3, geodesic segments in $\mathcal{B}$ depend uniquely and continuously on their endpoint pairs. Thus there is a subinterval $\mathbb{I}$ of [0, 1], that contains a neighborhood of t in [0, 1] and such that the arc $\gamma_n|_{\mathbb{I}}$ is minimizing for all large n. It follows that $\gamma_\infty|_{\mathbb{I}}$ is a geodesic, and therefore γ_∞ is a local geodesic.

The CAT(1) case is done in the same way, but one has to assume in addition that $R < \pi$. □

The following lemma and its proof were suggested to us by Alexander Lytchak. This lemma allows a local geodesic path to be moved continuously so that its endpoints follow given trajectories. This statement was originally proved by the first author and Richard Bishop using a different method; see [2].

3.2.2. Patchwork along a curve. *Let $\mathcal{U}$ be a proper length, locally* CAT(0) *space, and $\gamma\colon [0,1] \to \mathcal{U}$ be a path.*

Then there is a proper length CAT(0) *space $\mathcal{N}$, an open set $\hat{\Omega} \subset \mathcal{N}$, and a path $\hat{\gamma}\colon [0,1] \to \hat{\Omega}$, such that there is an open locally isometric immersion $\Phi\colon \hat{\Omega} \looparrowright \mathcal{U}$ satisfying $\Phi \circ \hat{\gamma} = \gamma$.*

If $\operatorname{length}\gamma < \pi$, *then the same holds in the* CAT(1) *case. Namely we assume that $\mathcal{U}$ is a proper length, locally* CAT(1) *space and construct a proper length* CAT(1) *space $\mathcal{N}$ with the same property as above.*

Proof. Fix $r > 0$ so that for each $t \in [0, 1]$, the closed ball $\overline{\mathrm{B}}[\gamma(t), r]$ forms a proper length CAT(κ) space.

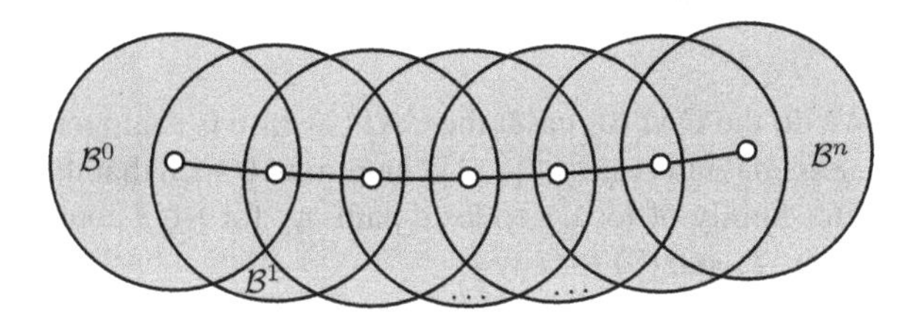

Choose a partition $0 = t^0 < t^1 < \cdots < t^n = 1$ such that

$$\mathrm{B}(\gamma(t^i), r) \supset \gamma([t^{i-1}, t^i])$$

for all $n > i > 0$. Set $\mathcal{B}^i = \overline{\mathrm{B}}[\gamma(t^i), r]$.

Consider the disjoint union $\bigsqcup_i \mathcal{B}^i = \{ (i, x) : x \in \mathcal{B}^i \}$ with the minimal equivalence relation $\sim$ such that $(i, x) \sim (i-1, x)$ for all i. Let $\mathcal{N}$ be the space obtained by gluing the $\mathcal{B}^i$ along $\sim$.

Note that $A^i = \mathcal{B}^i \cap \mathcal{B}^{i-1}$ is convex in $\mathcal{B}^i$ and in $\mathcal{B}^{i-1}$. Applying the Reshetnyak gluing theorem (2.3.1) n times, we conclude that $\mathcal{N}$ is a proper length CAT(0) space.

For $t \in [t^{i-1}, t^i]$, define $\hat{\gamma}(t)$ as the equivalence class of $(i, \gamma(t))$ in $\mathcal{N}$. Let $\hat{\Omega}$ be the ε-neighborhood of $\hat{\gamma}$ in $\mathcal{N}$, where $\varepsilon > 0$ is chosen so that $\mathrm{B}(\gamma(t), \varepsilon) \subset \mathcal{B}^i$ for all $t \in [t^{i-1}, t^i]$.

Define $\Phi\colon \hat{\Omega} \to \mathcal{U}$ by sending the equivalence class of (i, x) to x. It is straightforward to check that Φ, $\hat{\gamma}$, and $\hat{\Omega} \subset \mathcal{N}$ satisfy the conclusion of the lemma.

The CAT(1) case is proved in the same way. □

The following two corollaries follow from: (1) patchwork (3.2.2); (2) Proposition 2.2.7, which states that local geodesics are geodesics in any CAT(0) space; and (3) Theorem 2.2.3 on uniqueness of geodesics.

3.2.3. Corollary. *If $\mathcal{U}$ is a proper length, locally* CAT(0) *space, then for any pair of points $p, q \in \mathcal{U}$, the space of all local geodesic paths from p to q is discrete; that is, for any local geodesic path γ connecting p to q, there is $\varepsilon > 0$ such that for any other local geodesic path δ from p to q we have $|\gamma(t) - \delta(t)|_{\mathcal{U}} > \varepsilon$ for some $t \in [0, 1]$.*

Analogously, if $\mathcal{U}$ is a proper length, locally CAT(1) *space, then for any pair of points $p, q \in \mathcal{U}$, the space of all local geodesic paths shorter than π from p to q is discrete.*

3.2.4. Corollary. *If $\mathcal{U}$ is a proper length, locally* CAT(0) *space, then for any path α there is a choice of local geodesic path γ_α connecting the ends of α such that the map $\alpha \mapsto \gamma_\alpha$ is continuous, and if α is a local geodesic path then $\gamma_\alpha = \alpha$.*

Analogously, if $\mathcal{U}$ is a proper length, locally CAT(1) *space, then for any path α shorter than π, there is a choice of local geodesic path γ_α shorter than π connecting the ends of α such that the map $\alpha \mapsto \gamma_\alpha$ is continuous, and if α is a local geodesic path then $\gamma_\alpha = \alpha$.*

Proof of 3.2.4. We do the CAT(0) case; the CAT(1) case is analogous.

Consider the maximal interval $\mathbb{I} \subset [0, 1]$ containing 0 such that there is a continuous one-parameter family of local geodesic paths γ_t for $t \in \mathbb{I}$ connecting $\alpha(0)$ to $\alpha(t)$, with $\gamma_t(0) = \gamma_0(t) = \alpha(0)$ for any t.

By Proposition 3.2.1, $\mathbb{I}$ is closed, so we may assume $\mathbb{I} = [0, s]$ for some $s \in [0, 1]$.

Applying patchwork (3.2.2) to γ_s, we find that $\mathbb{I}$ is also open in $[0, 1]$. Hence $\mathbb{I} = [0, 1]$. Set $\gamma_\alpha = \gamma_1$.

By construction, if α is a local geodesic path, then $\gamma_\alpha = \alpha$.

Moreover, from Corollary 3.2.3, the construction $\alpha \mapsto \gamma_\alpha$ produces close results for sufficiently close paths in the metric defined by ❶; that is, the map $\alpha \mapsto \gamma_\alpha$ is continuous. □

Given a path $\alpha \colon [0, 1] \to \mathcal{U}$, we denote by $\bar{\alpha}$ the same path traveled in the opposite direction; that is,

$$\bar{\alpha}(t) = \alpha(1 - t).$$

The *product* of two paths will be denoted with "$*$"; if two paths α and β connect the same pair of points, then the product $\bar{\alpha} * \beta$ is a closed curve.

3.2.5. Exercise. *Assume $\mathcal{U}$ is a proper length, locally* CAT(1) *space. Consider the construction $\alpha \mapsto \gamma_\alpha$ provided by Corollary 3.2.4.*

Assume that α and β are two paths connecting the same pair of points in $\mathcal{U}$, where each is shorter than π and the product $\bar{\alpha} * \beta$ is null-homotopic in the class of closed curves shorter than $2 \cdot \pi$. Show that $\gamma_\alpha = \gamma_\beta$.

3.3 Globalization

Riemannian manifolds with nonpositive sectional curvature are locally CAT(0). The original formulation of the *globalization theorem*, or *Hadamard–Cartan theorem*, states that if M is a complete Riemannian manifold with sectional curvature at most 0, then the exponential map at any point $p \in M$ is a covering; in particular it implies that the universal cover of M is diffeomorphic to the Euclidean space of the same dimension.

In this generality, this theorem appeared in the lectures of Elie Cartan; see [26]. This theorem was proved for surfaces in Euclidean 3-space by Hans von Mangoldt [49] and a few years later independently for two-dimensional Riemannian manifolds by Jacques Hadamard [39].

Formulations for metric spaces of different generality were proved by Herbert Busemann in [24], Willi Rinow in [60], Mikhael Gromov in [36, p. 119]. A detailed proof of Gromov's statement was given by Werner Ballmann in [12] when $\mathcal{U}$ is proper, and by the first author and Richard Bishop in [2] in more generality; also see references in our book [6].

For proper CAT(1) spaces, the globalization theorem was proved by Brian Bowditch in [17].

3.3.1. Globalization theorem. *If a proper length, locally* CAT(0) *space is simply connected, then it is* CAT(0).

Analogously, suppose $\mathcal{U}$ is a proper length, locally CAT(1) *space such that any closed curve $\gamma\colon \mathbb{S}^1 \to \mathcal{U}$ shorter than $2 \cdot \pi$ is null-homotopic in the class of closed curves shorter than $2 \cdot \pi$. Then $\mathcal{U}$ is* CAT(1).

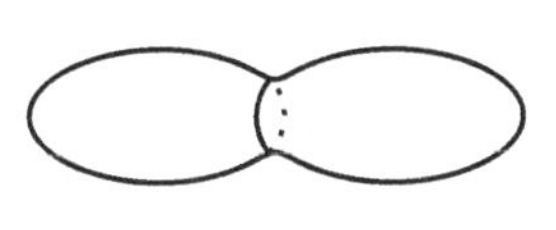

The surface on the diagram is an example of a simply connected space that is locally CAT(1) but not CAT(1). To contract the marked curve one has to increase its length to $2 \cdot \pi$ or more; in particular the surface does not satisfy the assumption of the globalization theorem.

The proof of the globalization theorem relies on the following theorem, which is essentially [10, Satz 9].

3.3.2. Patchwork globalization theorem. *A proper length, locally* CAT(0) *space $\mathcal{U}$ is* CAT(0) *if and only if all pairs of points in $\mathcal{U}$ are joined by unique geodesics, and these geodesics depend continuously on their endpoint pairs.*

Analogously, a proper length, locally CAT(1) *space $\mathcal{U}$ is* CAT(1) *if and only if all pairs of points in $\mathcal{U}$ at distance less than π are joined by unique geodesics, and these geodesics depend continuously on their endpoint pairs.*

The proof uses a thin-triangle decomposition with the inheritance lemma (2.2.9) and the following construction:

3.3.3. Line-of-sight map. *Let p be a point and α be a curve of finite length in a length space $\mathcal{X}$. Let $\mathring{\alpha} : [0, 1] \to \mathcal{U}$ be the constant-speed parametrization of α. If $\gamma_t\colon [0, 1] \to \mathcal{U}$ is a geodesic path from p to $\mathring{\alpha}(t)$, we say*

$$[0, 1] \times [0, 1] \to \mathcal{U}\colon (t, s) \mapsto \gamma_t(s)$$

is a line-of-sight map from p to α.

Proof of the patchwork globalization theorem (3.3.2). Note that the implication "only if" is already proved in Theorem 2.2.3; it only remains to prove the "if" part.

Fix a triangle $[pxy]$ in $\mathcal{U}$. We need to show that $[pxy]$ is thin.

By the assumptions, the line-of-sight map $(t, s) \mapsto \gamma_t(s)$ from p to $[xy]$ is uniquely defined and continuous.

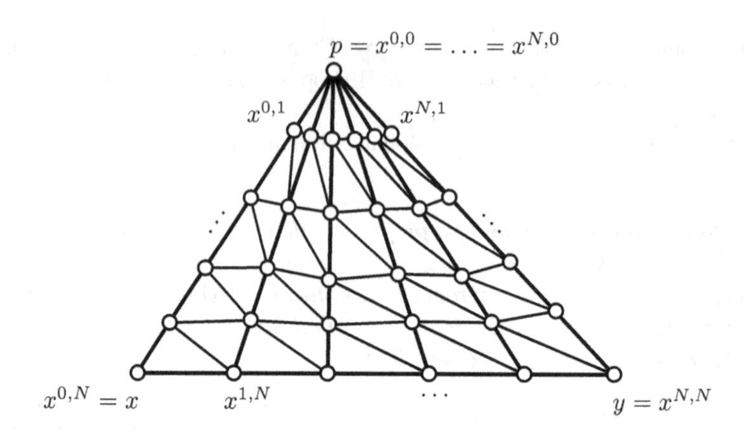

Fix a partition

$$0 = t^0 < t^1 < \cdots < t^N = 1,$$

and set $x^{i,j} = \gamma_{t^i}(t^j)$. Since the line-of-sight map is continuous and $\mathcal{U}$ is locally CAT(0), we may assume that the triangles

$$[x^{i,j} x^{i,j+1} x^{i+1,j+1}] \quad \text{and} \quad [x^{i,j} x^{i+1,j} x^{i+1,j+1}]$$

are thin for each pair i, j.

Now we show that the thin property propagates to $[pxy]$ by repeated application of the inheritance lemma (2.2.9):

- For fixed i, sequentially applying the lemma shows that the triangles $[px^{i,1}x^{i+1,2}]$, $[px^{i,2}x^{i+1,2}]$, $[px^{i,2}x^{i+1,3}]$, and so on are thin.

In particular, for each i, the long triangle $[px^{i,N}x^{i+1,N}]$ is thin.

- Applying the same lemma again shows that the triangles $[px^{0,N}x^{2,N}]$, $[px^{0,N}x^{3,N}]$, and so on, are thin.

In particular, $[pxy] = [px^{0,N}x^{N,N}]$ is thin. □

Proof of the globalization theorem; CAT(0) *case.* Let $\mathcal{U}$ be a proper length, locally CAT(0) space that is simply connected. Given a path α in $\mathcal{U}$, denote by γ_α the local geodesic path provided by Corollary 3.2.4. Since the map $\alpha \mapsto \gamma_\alpha$ is continuous, by Corollary 3.2.3 we have $\gamma_\alpha = \gamma_\beta$ for any pair of paths α and β homotopic relative to the ends.

Since $\mathcal{U}$ is simply connected, any pair of paths with common ends are homotopic. In particular, if α and β are local geodesics from p to q, then $\alpha = \gamma_\alpha = \gamma_\beta = \beta$ by Corollary 3.2.4. It follows that any two points $p, q \in \mathcal{U}$ are joined by a unique local geodesic that depends continuously on (p, q).

Since $\mathcal{U}$ is geodesic, it remains to apply the patchwork globalization theorem (3.3.2).

CAT(1) *case.* The proof goes along the same lines, but one needs to use Exercise 3.2.5. □

3.3.4. Corollary. *Any compact length, locally* CAT(0) *space that contains no closed local geodesics is* CAT(0).

Analogously, any compact length, locally CAT(1) *space that contains no closed local geodesics shorter than* $2 \cdot \pi$ *is* CAT(1).

Proof. By the globalization theorem (3.3.1), we need to show that the space is simply connected. Assume the contrary. Fix a nontrivial homotopy class of closed curves.

Denote by ℓ the exact lower bound for the lengths of curves in the class. Note that $\ell > 0$; otherwise there would be a closed noncontractible curve in a CAT(0) neighborhood of some point, contradicting Corollary 2.2.6.

Since the space is compact, the class contains a length-minimizing curve, which must be a closed local geodesic.

The CAT(1) case is analogous, but one has to consider a homotopy class of closed curves shorter than $2 \cdot \pi$. □

3.3.5. Exercise. *Prove that any compact length, locally* CAT(0) *space* $\mathcal{X}$ *that is not* CAT(0) *contains a* geodesic circle*; that is, a simple closed curve* γ *such that for any two points* $p, q \in \gamma$*, one of the arcs of* γ *with endpoints* p *and* q *is a geodesic.*

Formulate and prove the analogous statement for CAT(1) *spaces.*

3.3.6. Exercise. *Let* $\mathcal{U}$ *be a proper length* CAT(0) *space. Assume* $\tilde{\mathcal{U}} \to \mathcal{U}$ *is a metric double cover branching along a geodesic. Show that* $\tilde{\mathcal{U}}$ *is* CAT(0).

3.4 Polyhedral spaces

3.4.1. Definition. *A length space* $\mathcal{P}$ *is called a* (spherical) polyhedral space *if it admits a finite triangulation* τ *such that every simplex in* τ *is isometric to a simplex in a Euclidean space (or correspondingly a unit sphere) of appropriate dimension.*

By a triangulation of a polyhedral space *we will always understand a triangulation as above.*

Note that according to the above definition, all polyhedral spaces are compact. However, most of the statements below admit straightforward generalizations to *locally polyhedral spaces*; that is, complete length spaces, any point of which admits a closed neighborhood isometric to a polyhedral space. The latter class of spaces includes in particular infinite covers of polyhedral spaces.

The *dimension* of a polyhedral space $\mathcal{P}$ is defined as the maximal dimension of the simplices in one (and therefore any) triangulation of $\mathcal{P}$.

Links. Let $\mathcal{P}$ be a polyhedral space and σ be a simplex in a triangulation τ of $\mathcal{P}$.

The simplices that contain σ form an abstract simplicial complex called the *link* of σ, denoted by Link_σ. If m is the dimension of σ, then the set of vertices of Link_σ is formed by the $(m+1)$-simplices that contain σ; the set of its edges are formed by the $(m+2)$-simplices that contain σ; and so on.

The link Link_σ can be identified with the subcomplex of τ formed by all the simplices σ' such that $\sigma \cap \sigma' = \varnothing$ but both σ and σ' are faces of a simplex of τ.

The points in Link_σ can be identified with the normal directions to σ at a point in its interior. The angle metric between directions makes Link_σ into a spherical polyhedral space. We will always consider the link with this metric.

Tangent space and space of directions. Let $\mathcal{P}$ be a polyhedral space (Euclidean or spherical) and τ be its triangulation. If a point $p \in \mathcal{P}$ lies in the interior of a k-simplex σ of τ, then the tangent space $\mathrm{T}_p = \mathrm{T}_p\mathcal{P}$ is naturally isometric to

$$\mathbb{E}^k \times (\mathrm{Cone}\,\mathrm{Link}_\sigma).$$

Equivalently, the space of directions $\Sigma_p = \Sigma_p\mathcal{P}$ can be isometrically identified with the k-times iterated suspension over Link_σ; that is,

$$\Sigma_p \overset{iso}{=\!=} \mathrm{Susp}^k(\mathrm{Link}_\sigma).$$

If $\mathcal{P}$ is an m-dimensional polyhedral space, then for any $p \in \mathcal{P}$ the space of directions Σ_p is a spherical polyhedral space of dimension at most $m - 1$.

In particular, for any point p in σ, the isometry class of Link_σ together with $k = \dim \sigma$ determines the isometry class of Σ_p, and the other way around: Σ_p and k determine the isometry class of Link_σ.

A small neighborhood of p is isometric to a neighborhood of the tip of $\mathrm{Cone}\,\Sigma_p$. (If $\mathcal{P}$ is a spherical polyhedral space, then a small neighborhood of p is isometric to a neighborhood of the north pole in $\mathrm{Susp}\,\Sigma_p$.) In fact, if this property holds at any point of a compact length space $\mathcal{P}$, then $\mathcal{P}$ is a polyhedral space; see [45] by Nina Lebedeva and the third author.

The following theorem provides a combinatorial description of polyhedral spaces with curvature bounded above.

3.4.2. Theorem. *Let $\mathcal{P}$ be a polyhedral space and τ be its triangulation. Then $\mathcal{P}$ is locally* CAT(0) *if and only if the link of each simplex in τ has no closed local geodesic shorter than $2 \cdot \pi$.*

Analogously, let $\mathcal{P}$ be a spherical polyhedral space and τ be its triangulation. Then $\mathcal{P}$ is CAT(1) *if and only if neither $\mathcal{P}$ nor the link of any simplex in τ has a closed local geodesic shorter than $2 \cdot \pi$.*

Proof of 3.4.2. The "only if" part follows from Proposition 2.2.7 and Exercise 2.1.2.

To prove the "if" part, we apply induction on $\dim \mathcal{P}$. The base case $\dim \mathcal{P} = 0$ is evident. Let us start with the CAT(1) case.

Induction Step. Assume that the theorem is proved in the case $\dim \mathcal{P} < m$. Suppose $\dim \mathcal{P} = m$.

Fix a point $p \in \mathcal{P}$. A neighborhood of p is isometric to a neighborhood of the north pole in the suspension over the space of directions Σ_p.

Note that Σ_p is a spherical polyhedral space, and its links are isometric to links of $\mathcal{P}$. By the induction hypothesis, Σ_p is CAT(1). Thus, by the second part of Exercise 2.1.2, $\mathcal{P}$ is locally CAT(1).

Applying the second part of Corollary 3.3.4, we get the statement.

The CAT(0) case is done in exactly the same way except we need to use the first part of Exercise 2.1.2 and the first part of Corollary 3.3.4 on the last step. □

3.4.3. Exercise. *Let $\mathcal{P}$ be a polyhedral space such that any two points can be connected by a unique geodesic. Show that $\mathcal{P}$ is* CAT(0).

3.4.4. Advanced exercise. *Construct a Euclidean polyhedral metric on $\mathbb{S}^3$ such that the total angle around each edge in its triangulation is at least $2 \cdot \pi$.*

3.5 Flag complexes

3.5.1. Definition. *A simplicial complex $\mathcal{S}$ is called* flag *if whenever $\{v^0, \dots, v^k\}$ is a set of distinct vertices of $\mathcal{S}$ that are pairwise joined by edges, then the vertices $v^0, \dots, v^k$ span a k-simplex in $\mathcal{S}$.*

If the above condition is satisfied for $k = 2$, then we say that $\mathcal{S}$ satisfies the no-triangle condition.

Note that every flag complex is determined by its one-skeleton. Moreover, for any graph, its *cliques* (that is, complete subgraphs) define a flag complex. For that reason flag complexes are also called *clique complexes*.

3.5.2. Exercise. *Show that the barycentric subdivision of any simplicial complex is a flag complex.*

Use the flag condition (see 3.5.5 *below) to conclude that any finite simplicial complex is homeomorphic to a proper length* CAT(1) *space.*

3.5.3. Proposition. *A simplicial complex $\mathcal{S}$ is flag if and only if $\mathcal{S}$ as well as all the links of all its simplices satisfy the no-triangle condition.*

From the definition of flag complex we get the following.

3.5.4. Observation. *Any link of any simplex in a flag complex is flag.*

Proof of 3.5.3. By Observation 3.5.4, the no-triangle condition holds for any flag complex and the links of all its simplices.

Now assume that a complex $\mathcal{S}$ and all its links satisfy the no-triangle condition. It follows that $\mathcal{S}$ includes a 2-simplex for each triangle. Applying the same observation for each edge we get that $\mathcal{S}$ includes a 3-simplex for any complete graph with 4 vertices. Repeating this observation for triangles, 4-simplices, 5-simplices, and so on, we get that $\mathcal{S}$ is flag. □

All-right triangulation. A triangulation of a spherical polyhedral space is called an *all-right triangulation* if each simplex of the triangulation is isometric to a spherical simplex all of whose angles are right. Similarly, we say that a simplicial complex is equipped with an *all-right spherical metric* if it is a length metric and each simplex is isometric to a spherical simplex all of whose angles are right.

Spherical polyhedral CAT(1) spaces glued from right-angled simplices admit the following characterization discovered by Mikhael Gromov [36, p. 122].

3.5.5. Flag condition. *Assume that a spherical polyhedral space $\mathcal{P}$ admits an all-right triangulation τ. Then $\mathcal{P}$ is* CAT(1) *if and only if τ is flag.*

Proof. *"only if" part.* Assume there are three vertices v^1, v^2, and v^3 of τ that are pairwise joined by edges but do not span a triangle. Note that in this case

$$\measuredangle[v^1\,{}^{v^2}_{v^3}] = \measuredangle[v^2\,{}^{v^3}_{v^1}] = \measuredangle[v^3\,{}^{v^1}_{v^2}] = \pi.$$

Equivalently,

❶ *The product of the geodesics $[v^1v^2]$, $[v^2v^3]$, and $[v^3v^1]$ forms a locally geodesic loop in $\mathcal{P}$ of length $\frac{3}{2}\cdot\pi$.*

Now assume that $\mathcal{P}$ is CAT(1). Then by Theorem 3.4.2, $\mathrm{Link}_\sigma\,\mathcal{P}$ is CAT(1) for every simplex σ in τ.

Each of these links is an all-right spherical complex and by Theorem 3.4.2, none of these links can contain a geodesic circle shorter than $2\cdot\pi$.

Therefore Proposition 3.5.3 and ❶ imply the "only if" part.

"If" part. By Observation 3.5.4 and Theorem 3.4.2, it is sufficient to show that any closed local geodesic γ in a flag complex $\mathcal{S}$ with all-right metric has length at least $2\cdot\pi$.

Recall that the *closed star* of a vertex v (briefly $\overline{\mathrm{Star}}_v$) is formed by all the simplices containing v. Similarly, Star_v, the open star of v, is the union of all simplices containing v with faces opposite v removed.

Choose a vertex v such that Star_v contains a point $\gamma(t_0)$ of γ. Consider the maximal arc γ_v of γ that contains the point $\gamma(t_0)$ and runs in Star_v. Note that the distance $|v-\gamma_v(t)|_{\mathcal{P}}$ behaves in exactly the same way as the distance from the north pole in $\mathbb{S}^2$ to a geodesic in the north hemisphere; that is, there is a geodesic $\tilde\gamma_v$ in the north hemisphere of $\mathbb{S}^2$ such that for any t we have

$$|v-\gamma_v(t)|_{\mathcal{P}} = |n-\tilde\gamma_v(t)|_{\mathbb{S}^2},$$

where n denotes the north pole of $\mathbb{S}^2$. In particular,

$$\mathrm{length}\,\gamma_v = \pi;$$

that is, γ spends time π on every visit to Star_v.

After leaving Star_v, the local geodesic γ has to enter another simplex, say σ'. Since τ is flag, the simplex σ' has a vertex v' not joined to v by an edge; that is,

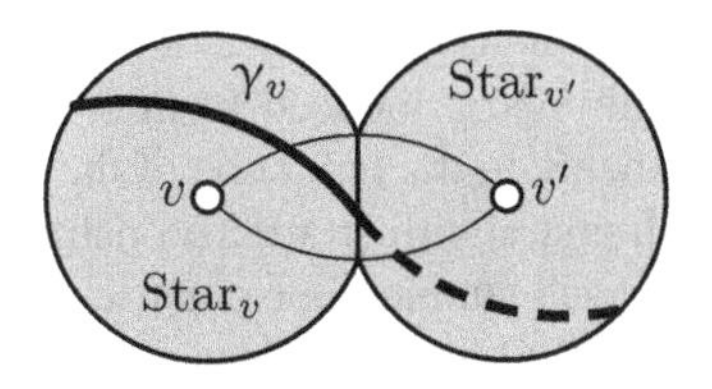

$$\mathrm{Star}_v \cap \mathrm{Star}_{v'} = \varnothing$$

The same argument as above shows that γ spends time π on every visit to $\mathrm{Star}_{v'}$. Therefore the total length of γ is at least $2 \cdot \pi$. □

3.5.6. Exercise. *Assume that a spherical polyhedral space $\mathcal{P}$ admits a triangulation τ such that all edge lengths of all simplices are at least $\frac{\pi}{2}$. Show that $\mathcal{P}$ is* CAT(1) *if τ is flag.*

The space of trees. The following construction is given by Louis Billera, Susan Holmes, and Karen Vogtmann in [14].

Let $\mathcal{T}_n$ be the set of all metric trees with n end vertices labeled by $a^1, \dots, a^n$. To describe one tree in $\mathcal{T}_n$ we may fix a topological trivalent tree t with end vertices $a^1, \dots, a^n$ and all other vertices of degree 3 and prescribe the lengths of $2 \cdot n - 3$ edges. If the length of an edge vanishes, we assume that this edge degenerates; such a tree can be also described using a different topological tree t'. The subset of $\mathcal{T}_n$ corresponding to the given topological tree t can be identified with the octant

$$\left\{ (x_1, \dots, x_{2\cdot n-3}) \in \mathbb{R}^{2\cdot n-3} : x_i \geqslant 0 \right\}.$$

Equip each such subset with the metric induced from $\mathbb{R}^{2\cdot n-3}$ and consider the length metric on $\mathcal{T}_n$ induced by these metrics.

3.5.7. Exercise. *Show that $\mathcal{T}_n$ with the described metric is* CAT(0).

3.6 Cubical complexes

The definition of a cubical complex mostly repeats the definition of a simplicial complex, with simplices replaced by cubes.

Formally, a *cubical complex* is defined as a subcomplex of the unit cube in the Euclidean space $\mathbb{R}^N$ of large dimension; that is, a collection of faces of the cube such that together with each face it contains all its subfaces. Each cube face in this collection will be called a *cube* of the cubical complex.

Note that according to this definition, any cubical complex is finite.

The union of all the cubes in a cubical complex $\mathcal{Q}$ will be called its *underlying space*. A homeomorphism from the underlying space of $\mathcal{Q}$ to a topological space $\mathcal{X}$ is called a *cubulation of $\mathcal{X}$*.

The underlying space of a cubical complex $\mathcal{Q}$ will be always considered with the length metric induced from $\mathbb{R}^N$. In particular, with this metric, each cube of $\mathcal{Q}$ is isometric to the unit cube of the corresponding dimension.

It is straightforward to construct a triangulation of the underlying space of $\mathcal{Q}$ such that each simplex is isometric to a Euclidean simplex. In particular the underlying space of $\mathcal{Q}$ is a Euclidean polyhedral space.

The link of a cube in a cubical complex is defined similarly to the link of a simplex in a simplicial complex. It is a simplicial complex that admits a natural all-right triangulation—each simplex corresponds to an adjusted cube.

Cubical analog of a simplicial complex. Let $\mathcal{S}$ be a finite simplicial complex and $\{v_1, \dots, v_N\}$ be the set of its vertices.

Consider $\mathbb{R}^N$ with the standard basis $\{e_1, \dots, e_N\}$. Denote by $\square^N$ the standard unit cube in $\mathbb{R}^N$; that is,

$$\square^N = \left\{ (x_1, \dots, x_N) \in \mathbb{R}^N : 0 \leqslant x_i \leqslant 1 \text{ for each } i \right\}.$$

Given a k-dimensional simplex $\langle v_{i_0}, \dots, v_{i_k}\rangle$ in $\mathcal{S}$, mark the $(k+1)$-dimensional faces in $\square^N$ (there are 2^{N-k} of them) which are parallel to the coordinate $(k+1)$-plane spanned by $e_{i_0}, \dots, e_{i_k}$.

Note that the set of all marked faces of $\square^N$ forms a cubical complex; it will be called the *cubical analog* of $\mathcal{S}$ and will be denoted as $\square_{\mathcal{S}}$.

3.6.1. Proposition. *Let $\mathcal{S}$ be a finite connected simplicial complex and $\mathcal{Q} = \square_{\mathcal{S}}$ be its cubical analog. Then the underlying space of $\mathcal{Q}$ is connected and the link of any vertex of $\mathcal{Q}$ is isometric to $\mathcal{S}$ equipped with the spherical right-angled metric.*

In particular, if $\mathcal{S}$ is a flag complex, then $\mathcal{Q}$ is a locally CAT(0) *and therefore its universal cover $\tilde{\mathcal{Q}}$ is* CAT(0).

Proof. The first part of the proposition follows from the construction of $\square_{\mathcal{S}}$.

If $\mathcal{S}$ is flag, then by the flag condition (3.5.5) the link of any cube in $\mathcal{Q}$ is CAT(1). Therefore, by the cone construction (Exercise 2.1.2) $\mathcal{Q}$ is locally CAT(0). It remains to apply the globalization theorem (3.3.1). □

From Proposition 3.6.1, it follows that the cubical analog of any flag complex is aspherical. The following exercise states that the converse also holds; see [31, 5.4].

3.6.2. Exercise. *Show that a finite simplicial complex is flag if and only if its cubical analog is aspherical.*

3.7 Exotic aspherical manifolds

By the globalization theorem (3.3.1), any proper length CAT(0) space is contractible. Therefore all proper length, locally CAT(0) spaces, are *aspherical*; that is, they have contractible universal covers. This observation can be used to construct examples of aspherical spaces.

Let $\mathcal{X}$ be a proper topological space. Recall that $\mathcal{X}$ is called *simply connected at infinity* if for any compact set $K \subset \mathcal{X}$ there is a bigger compact set $K' \supset K$ such that $\mathcal{X}\backslash K'$ is path connected and any loop which lies in $\mathcal{X}\backslash K'$ is null-homotopic in $\mathcal{X}\backslash K$.

Recall that path connected spaces are not empty by definition. Therefore compact spaces are not simply connected at infinity.

The following example was constructed by Michael Davis in [30].

3.7.1. Proposition. *For any $m \geqslant 4$ there is a closed aspherical m-dimensional manifold whose universal cover is not simply connected at infinity.*

In particular, the universal cover of this manifold is not homeomorphic to the m-dimensional Euclidean space.

The proof requires the following lemma.

3.7.2. Lemma. *Let $\mathcal{S}$ be a finite flag complex, $\mathcal{Q} = \square_{\mathcal{S}}$ be its cubical analog and $\tilde{\mathcal{Q}}$ be the universal cover of $\mathcal{Q}$.*

Assume $\tilde{\mathcal{Q}}$ is simply connected at infinity. Then $\mathcal{S}$ is simply connected.

Proof. Assume $\mathcal{S}$ is not simply connected. Equip $\mathcal{S}$ with an all-right spherical metric. Choose a shortest noncontractible circle $\gamma\colon \mathbb{S}^1 \to \mathcal{S}$ formed by the edges of $\mathcal{S}$.

Note that γ forms a one-dimensional subcomplex of $\mathcal{S}$ which is a closed local geodesic. Denote by G the subcomplex of $\mathcal{Q}$ which corresponds to γ.

Fix a vertex $v \in G$; let G_v be the connected component of v in G. Let $\tilde{G}$ be a connected component of the inverse image of G_v in $\tilde{\mathcal{Q}}$ for the universal cover $\tilde{\mathcal{Q}} \to \mathcal{Q}$. Fix a point $\tilde{v} \in \tilde{G}$ in the inverse image of v.

Note that

❶ *$\tilde{G}$ is a convex set in $\tilde{\mathcal{Q}}$.*

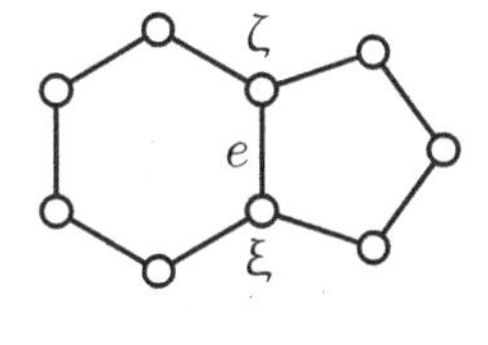

Indeed, according to Proposition 3.6.1, $\tilde{\mathcal{Q}}$ is CAT(0). By Exercise 2.2.12, it is sufficient to show that $\tilde{G}$ is locally convex in $\tilde{\mathcal{Q}}$, or equivalently, G is locally convex in $\mathcal{Q}$.

Note that the latter can only fail if γ contains two vertices, say ξ and ζ in $\mathcal{S}$, which are joined by an edge not in γ; denote this edge by e.

Each edge of $\mathcal{S}$ has length $\frac{\pi}{2}$. Therefore each of two circles formed by e and an arc of γ from ξ to ζ is shorter that γ. Moreover, at least one of them is noncontractible since γ is noncontractible. That is, γ is not a shortest noncontractible circle, a contradiction. △

Further, note that $\tilde{G}$ is homeomorphic to the plane, since $\tilde{G}$ is a two-dimensional manifold without boundary which by the above is CAT(0) and hence is contractible.

Denote by C_R the circle of radius R in $\tilde{G}$ centered at $\tilde{v}$. All C_R are homotopic to each other in $\tilde{G}\backslash\{\tilde{v}\}$ and therefore in $\tilde{\mathcal{Q}}\backslash\{\tilde{v}\}$.

Note that the map $\tilde{\mathcal{Q}}\backslash\{\tilde{v}\} \to \mathcal{S}$ which returns the direction of $[\tilde{v}x]$ for any $x \neq \tilde{v}$, maps C_R to a circle homotopic to γ. Therefore C_R is not contractible in $\tilde{\mathcal{Q}}\backslash\{\tilde{v}\}$.

If R is large, the circle C_R lies outside of any fixed compact set K' in $\tilde{\mathcal{Q}}$. From above C_R is not contractible in $\tilde{\mathcal{Q}}\backslash K$ if $K \supset \tilde{v}$. It follows that $\tilde{\mathcal{Q}}$ is not simply connected at infinity, a contradiction. □

The proof of the following exercise is analogous. It will be used later in the proof of Proposition 3.7.4—a more geometric version of Proposition 3.7.1.

3.7.3. Exercise. *Under the assumptions of Lemma 3.7.2, for any vertex v in $\mathcal{S}$ the complement $\mathcal{S}\backslash\{v\}$ is simply connected.*

Proof of 3.7.1. Let Σ^{m-1} be an $(m-1)$-dimensional smooth homology sphere that is not simply connected and bounds a contractible smooth compact m-dimensional manifold $\mathcal{W}$.

For $m \geqslant 5$ the existence of such $(\mathcal{W}, \Sigma)$ follows from [43]. For $m = 4$ it follows from the construction in [48].

Pick any triangulation τ of W and let $\mathcal{S}$ be the resulting subcomplex that triangulates Σ.

We can assume that $\mathcal{S}$ is flag; otherwise pass to the barycentric subdivision of τ and apply Exercise 3.5.2.

Let $\mathcal{Q} = \square_{\mathcal{S}}$ be the cubical analog of $\mathcal{S}$.

By Proposition 3.6.1, $\mathcal{Q}$ is a homology manifold. It follows that $\mathcal{Q}$ is a piecewise linear manifold with a finite number of singularities at its vertices.

Removing a small contractible neighborhood V_v of each vertex v in $\mathcal{Q}$, we can obtain a piecewise linear manifold $\mathcal{N}$ whose boundary is formed by several copies of Σ.

Let us glue a copy of $\mathcal{W}$ along its boundary to each copy of Σ in the boundary of $\mathcal{N}$. This results in a closed piecewise linear manifold $\mathcal{M}$ which is homotopically equivalent to $\mathcal{Q}$.

Indeed, since both V_v and $\mathcal{W}$ are contractible, the identity map of their common boundary Σ can be extended to a homotopy equivalence $V_v \to \mathcal{W}$ relative to the boundary. Therefore the identity map on $\mathcal{N}$ extends to homotopy equivalences $f\colon \mathcal{Q} \to \mathcal{M}$ and $g\colon \mathcal{M} \to \mathcal{Q}$.

Finally, by Lemma 3.7.2, the universal cover $\tilde{\mathcal{Q}}$ of $\mathcal{Q}$ is not simply connected at infinity.

The same holds for the universal cover $\tilde{\mathcal{M}}$ of $\mathcal{M}$. The latter follows since the constructed homotopy equivalences $f\colon \mathcal{Q} \to \mathcal{M}$ and $g\colon \mathcal{M} \to \mathcal{Q}$ lift to *proper maps* $\tilde{f}\colon \tilde{\mathcal{Q}} \to \tilde{\mathcal{M}}$ and $\tilde{g}\colon \tilde{\mathcal{M}} \to \tilde{\mathcal{Q}}$; that is, for any compact sets $A \subset \tilde{\mathcal{Q}}$ and $B \subset \tilde{\mathcal{M}}$, the inverse images $\tilde{g}^{-1}(A)$ and $\tilde{f}^{-1}(B)$ are compact. □

The following proposition was proved by Fredric Ancel, Michael Davis, and Craig Guilbault in [11]; it could be considered as a more geometric version of Proposition 3.7.1.

3.7.4. Proposition. *Given $m \geqslant 5$, there is a Euclidean polyhedral space $\mathcal{P}$ such that:*

(a) $\mathcal{P}$ is homeomorphic to a closed m-dimensional manifold.

(b) $\mathcal{P}$ *is locally* CAT(0).
(c) The universal cover of $\mathcal{P}$ *is not simply connected at infinity.*

There are no three-dimensional examples of that type; see [59] by Dale Rolfsen. In [65], Paul Thurston conjectured that the same holds in the four-dimensional case.

Proof. Apply Exercise 3.7.3 to the barycentric subdivision of the simplicial complex $\mathcal{S}$ provided by Exercise 3.7.5. □

3.7.5. Exercise. *Given an integer* $m \geqslant 5$, *construct a finite* $(m-1)$*-dimensional simplicial complex* $\mathcal{S}$ *such that* $\operatorname{Cone}\mathcal{S}$ *is homeomorphic to* $\mathbb{E}^m$ *and* $\pi_1(\mathcal{S}\backslash\{v\}) \neq 0$ *for some vertex* v *in* $\mathcal{S}$.

3.8 Comments

As was mentioned earlier, the motivation for the notion of CAT(κ) spaces comes from the fact that a Riemannian manifold is locally CAT(κ) if and only if it has sec $\leqslant \kappa$. This easily follows from Rauch comparison for Jacobi fields and Proposition 2.2.2.

In the globalization theorem (3.3.1), properness can be weakened to completeness; see our book [6] and the references therein.

The condition on polyhedral CAT(κ) spaces given in Theorem 3.4.2 might look easy to use, but in fact, it is hard to check even in very simple cases. For example the description of those coverings of $\mathbb{S}^3$ branching at three great circles which are CAT(1) requires quite a bit of work; see [27]—try to guess the answer before reading.

Another example is the space $\mathcal{B}_4$ that is the universal cover of $\mathbb{C}^4$ infinitely branching in six complex planes $z_i = z_j$ with the induced length metric. So far it is not known if $\mathcal{B}_4$ is CAT(0). Understanding this space could be helpful for studying the braid group on 4 strings; read [52] by Dmitri Panov and the third author for more on it. This circle of questions is closely related to the generalization of the flag condition (3.5.5) to spherical simplices with few acute dihedral angles.

The construction used in the proof of Proposition 3.7.1 admits a number of interesting modifications, several of which are discussed in the survey [31] by Michael Davis.

A similar argument was used by Michael Davis, Tadeusz Januszkiewicz, and Jean-François Lafont in [33]. They constructed a closed smooth four-dimensional manifold M with universal cover $\tilde{M}$ diffeomorphic to $\mathbb{R}^4$, such that M admits a polyhedral metric which is locally CAT(0), but does not admit a Riemannian metric with nonpositive sectional curvature. Another example of that type was constructed by Stephan Stadler; see [63]. There are no lower dimensional examples of this type—the two-dimensional case follows from the classification of surfaces, and the three-dimensional case follows from the geometrization conjecture.

It is noteworthy that any complete, simply connected Riemannian manifold with nonpositive curvature is homeomorphic to the Euclidean space of the same dimension. In fact, by the globalization theorem (3.3.1), the exponential map at a point of such a manifold is a homeomorphism. In particular, there is no Riemannian analog of Proposition 3.7.4.

Recall that a triangulation of an m-dimensional manifold defines a piecewise linear structure if the link of every simplex Δ is homeomorphic to the sphere of dimension $m - 1 - \dim \Delta$. According to Stone's theorem, see [32, 64], the triangulation of $\mathcal{P}$ in Proposition 3.7.4 cannot be made piecewise linear—despite the fact that $\mathcal{P}$ is a manifold, its triangulation does not induce a piecewise linear structure.

The flag condition also leads to the so-called *hyperbolization* procedure, a flexible tool for constructing aspherical spaces; a good survey on the subject is given by Ruth Charney and Michael Davis in [28].

All the topics discussed in this chapter link Alexandrov geometry with the fundamental group. The theory of *hyperbolic groups*, a branch of *geometric group theory*, introduced by Mikhael Gromov [36], could be considered as a further step in this direction.

Chapter 4
Subsets

In this chapter we give a partial answer to the question:

Which subsets of Euclidean space, equipped with their induced length metrics, are CAT(0)*?*.

4.1 Motivating examples

Consider three subgraphs of different quadric surfaces:

$$A = \{ (x, y, z) \in \mathbb{E}^3 : z \leqslant x^2 + y^2 \},$$
$$B = \{ (x, y, z) \in \mathbb{E}^3 : z \leqslant -x^2 - y^2 \},$$
$$C = \{ (x, y, z) \in \mathbb{E}^3 : z \leqslant x^2 - y^2 \}.$$

4.1.1. Question. *Which of the sets A, B and C, if equipped with the induced length metric, are* CAT(0) *and why?*

The answers are given below, but it is instructive to think about these questions before reading further.

***A*.** No, A is not CAT(0).

The boundary ∂A is the paraboloid described by $z = x^2 + y^2$; in particular it bounds an open convex set in $\mathbb{E}^3$ whose complement is A. The closest-point projection of $A \to \partial A$ is short (Exercise 2.2.13). It follows that ∂A is a convex set in A equipped with its induced length metric.

Therefore if A is CAT(0), then so is ∂A. The latter is not true: ∂A is a smooth convex surface and has strictly positive curvature by the Gauss formula.

S. Alexander et al., *An Invitation to Alexandrov Geometry*,
SpringerBriefs in Mathematics, https://doi.org/10.1007/978-3-030-05312-3_4

***B*.** Yes, B is CAT(0).

Evidently B is a convex closed set in $\mathbb{E}^3$. Therefore the length metric on B coincides with the Euclidean metric and CAT(0) comparison holds.

***C*.** Yes, C is CAT(0), but the proof is not as easy as before. We give a sketch here; a complete proof of a more general statement is given in Section 4.3.

Set $f_t(x, y) = x^2 - y^2 - 2 \cdot (x - t)^2$. Consider the one-parameter family of sets

$$V_t = \{ (x, y, z) \in \mathbb{E}^3 : z \leqslant f_t(x, y) \}.$$

Each set V_t is a solid paraboloid tangent to ∂C along the parabola $y \mapsto (t, y, t^2 - y^2)$. The set V_t is closed and convex for any t, and

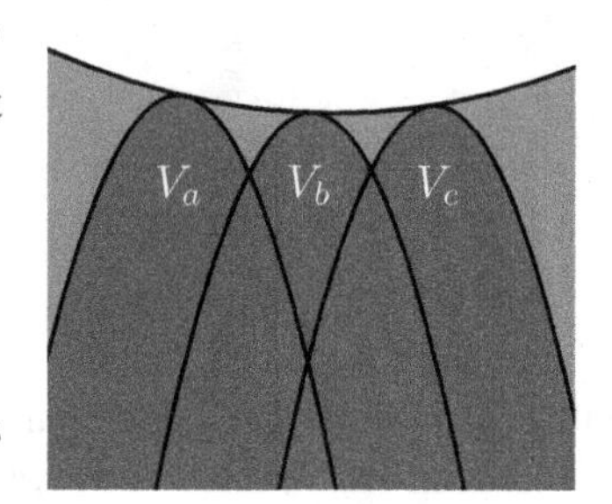

$$C = \bigcup_t V_t.$$

Further note that the function $t \mapsto f_t(x, y)$ is concave for any fixed x, y. Therefore

❶ *if $a < b < c$, then $V_b \supset V_a \cap V_c$.*

Consider the finite union

$$C' = V_{t_1} \cup \cdots \cup V_{t_n}.$$

The inclusion ❶ makes it possible to apply Reshetnyak gluing theorem 2.3.1 recursively and show that C' is CAT(0). By approximation, the CAT(0) comparison holds for any 4 points in C; hence C is CAT(0).

Remark. The set C is not convex, but it is *two-convex* as defined in the next section. As you will see, two-convexity is closely related to the inheritance of an upper curvature bound by a subset.

4.2 Two-convexity

The following definition is closely related to the one given by Mikhael Gromov in [38, §$\frac{1}{2}$], see also [51].

4.2.1. Definition. *We say that a subset $K \subset \mathbb{E}^m$ is* two-convex *if the following condition holds for any plane $W \subset \mathbb{E}^m$: If γ is a simple closed curve in $W \cap K$ that is null-homotopic in K, then it is null-homotopic in $W \cap K$, and in particular the disk in W bounded by γ lies in K.*

Note that two-convex sets do not have to be connected or simply connected. The following two propositions follow immediately from the definition.

4.2.2. Proposition. *Any subset in $\mathbb{E}^2$ is two-convex.*

4.2.3. Proposition. *The intersection of an arbitrary collection of two-convex sets in $\mathbb{E}^m$ is two-convex.*

4.2.4. Proposition. *The interior of any two-convex set in $\mathbb{E}^m$ is a two-convex set.*

Proof. Fix a two-convex set $K \subset \mathbb{E}^m$ and a 2-plane W; denote by $\operatorname{Int} K$ the interior of K. Let γ be a closed simple curve in $W \cap \operatorname{Int} K$ that is contractible in the interior of K.

Since K is two-convex, the plane disk D bounded by γ lies in K. The same holds for the translations of D by small vectors. Therefore D lies in $\operatorname{Int} K$; that is, $\operatorname{Int} K$ is two-convex. □

4.2.5. Definition. *Given a subset $K \subset \mathbb{E}^m$, define its two-convex hull (briefly,* $\operatorname{Conv}_2 K$*) as the intersection of all two-convex subsets containing K.*

Note that by Proposition 4.2.3, the two-convex hull of any set is two-convex. Further, by Proposition 4.2.4, the two-convex hull of an open set is open.

The next proposition describes closed two-convex sets with smooth boundary.

4.2.6. Proposition. *Let $K \subset \mathbb{E}^m$ be a closed subset.*

Assume that the boundary of K is a smooth hypersurface S. Consider the unit normal vector field ν on S that points outside of K. Denote by $k_1 \leqslant \dots \leqslant k_{m-1}$ the principal curvature functions of S with respect to ν (note that if K is convex, then $k_1 \geqslant 0$).

Then K is two-convex if and only if $k_2(p) \geqslant 0$ for any point $p \in S$. Moreover, if $k_2(p) < 0$ at some point p, then Definition 4.2.1 fails for some curve γ forming a triangle in an arbitrary small neighborhood of p.

The proof is taken from [38, §$\frac{1}{2}$], but we added some details.

Proof. "only if" part. If $k_2(p) < 0$ for some $p \in S$, consider the plane W containing p and spanned by the first two principal directions at p. Choose a small triangle in W which surrounds p and move it slightly in the direction of $\nu(p)$. We get a triangle $[xyz]$ which is null-homotopic in K, but the solid triangle $\Delta = \operatorname{Conv}\{x, y, x\}$ bounded by $[xyz]$ does not lie in K completely. Therefore K is not two-convex. (See figure in the "only if" part of the smooth two-convexity theorem (4.3.1)).

"If" part. Recall that a smooth function $f\colon \mathbb{E}^m \to \mathbb{R}$ is called *strongly convex* if its Hessian is positive definite at each point.

Suppose $f\colon \mathbb{E}^m \to \mathbb{R}$ is a smooth strongly convex function such that the restriction $f|_S$ is a Morse function. Note that a generic smooth strongly convex function $f\colon \mathbb{E}^m \to \mathbb{R}$ has this property.

For a critical point p of $f|_S$, the outer normal vector $\nu(p)$ is parallel to the gradient $\nabla_p f$; we say that p is a *positive critical point* if $\nu(p)$ and $\nabla_p f$ point in the same direction, and *negative* otherwise. If f is generic, then we can assume that the sign is defined for all critical points; that is, $\nabla_p f \neq 0$ for any critical point p of $f|_S$.

Since $k_2 \geqslant 0$ and the function f is strongly convex, the negative critical points of $f|_S$ have index at most 1.

Given a real value s, set

$$K_s = \{ x \in K : f(x) < s \}.$$

Assume $\varphi_0 \colon \mathbb{D} \to K$ is a continuous map of the disk $\mathbb{D}$ such that $\varphi_0(\partial\mathbb{D}) \subset K_s$.

Note that by the Morse lemma, there is a homotopy $\varphi_t : \mathbb{D} \to K$ rel $\partial\mathbb{D}$ such that $\varphi_1(\mathbb{D}) \subset K_s$.

Indeed, we can construct a homotopy $\varphi_t : \mathbb{D} \to K$ that decreases the maximum of $f \circ \varphi$ on $\mathbb{D}$ until the maximum occurs at a critical point p of $f|_S$. This point cannot be negative, otherwise its index would be at least 2. If this critical point is positive, then it is easy to decrease the maximum a little by pushing the disk from S into K in the direction of $-\nabla f_p$.

Consider a closed curve $\gamma \colon \mathbb{S}^1 \to K$ that is null-homotopic in K. Note that the distance function

$$f_0(x) = |\operatorname{Conv}\gamma - x|_{\mathbb{E}^m}$$

is convex. Therefore f_0 can be approximated by smooth strongly convex functions f in general position. From above, there is a disk in K with boundary γ that lies arbitrarily close to $\operatorname{Conv}\gamma$. Since K is closed, the statement follows. □

Note that the "if" part proves a somewhat stronger statement. Namely, any plane curve γ (*not necessary simple*) which is contractible in K is also contractible in the intersection of K with the plane of γ. The latter condition does not hold for the complement of two planes in $\mathbb{E}^4$, which is two-convex by Proposition 4.2.3; see also Exercise 4.5.3 below. The following proposition shows that there are no such examples in $\mathbb{E}^3$.

4.2.7. Proposition. *Let $\Omega \subset \mathbb{E}^3$ be an open two-convex subset. Then for any plane $W \subset \mathbb{E}^3$, any closed curve in $W \cap \Omega$ that is null-homotopic in Ω is also null-homotopic in $W \cap \Omega$.*

In the proof we use the following classical result:

4.2.8. Loop theorem. *Let M be a three-dimensional manifold with nonempty boundary ∂M. Assume $f \colon (\mathbb{D}, \partial\mathbb{D}) \to (M, \partial M)$ is a continuous map from the disk $\mathbb{D}$ such that the boundary curve $f|_{\partial\mathbb{D}}$ is not null-homotopic in ∂M. Then there is an embedding $h \colon (\mathbb{D}, \partial\mathbb{D}) \to (M, \partial M)$ with the same property.*

The theorem is due to Christos Papakyriakopoulos; a proof can be found in [40].

Proof of 4.2.7. Fix a closed plane curve γ in $W \cap \Omega$ that is null-homotopic in Ω. Suppose γ is not contractible in $W \cap \Omega$.

Let $\varphi \colon \mathbb{D} \to \Omega$ be a map of the disk with the boundary curve γ.

Since Ω is open we can first change φ slightly so that $\varphi(x) \notin W$ for $1 - \varepsilon < |x| < 1$ for some small $\varepsilon > 0$. By further changing φ slightly we can assume that it is transversal to W on $\operatorname{Int}\mathbb{D}$ and agrees with the previous map near $\partial\mathbb{D}$.

This means that $\varphi^{-1}(W) \cap \operatorname{Int}\mathbb{D}$ consists of finitely many simple closed curves which cut $\mathbb{D}$ into several components. Consider one of the "innermost" components c'; that is, c' is a boundary curve of a disk $\mathbb{D}' \subset \mathbb{D}$, $\varphi(c')$ is a closed curve in W and $\varphi(\mathbb{D}')$ completely lies in one of the two half-spaces with boundary W. Denote this half-space by H.

If $\varphi(c')$ is not contractible in $W \cap \Omega$, then applying the loop theorem to $M^3 = H \cap \Omega$ we conclude that there exists a *simple* closed curve $\gamma' \subset \Omega \cap W$ which is not contractible in $\Omega \cap W$ but is contractible in $\Omega \cap H$. This contradicts two-convexity of Ω.

Hence $\varphi(c')$ is contractible in $W \cap \Omega$. Therefore φ can be changed in a small neighborhood U of $\mathbb{D}'$ so that the new map $\hat{\varphi}$ maps U to one side of W. In particular, the set $\hat{\varphi}^{-1}(W)$ consists of the same curves as $\varphi^{-1}(W)$ with the exception of c'.

Repeating this process several times we reduce the problem to the case where $\varphi^{-1}(W) \cap \operatorname{Int}\mathbb{D} = \varnothing$. This means that $\varphi(\mathbb{D})$ lies entirely in one of the half-spaces bounded by W.

Again applying the loop theorem, we obtain a simple closed curve in $W \cap \Omega$ which is not contractible in $W \cap \Omega$ but is contractible in Ω. This again contradicts two-convexity of Ω. Hence γ is contractible in $W \cap \Omega$ as claimed. □

4.3 Sets with smooth boundary

In this section we characterize the subsets with smooth boundary in $\mathbb{E}^m$ that form CAT(0) spaces.

4.3.1. Smooth two-convexity theorem. *Let K be a closed, simply connected subset in $\mathbb{E}^m$ equipped with the induced length metric. Assume K is bounded by a smooth hypersurface. Then K is* CAT(0) *if and only if K is two-convex.*

This theorem is a baby case of the main result in [1], which is briefly discussed at the end of the chapter.

Proof. Denote by S and by Ω the boundary and the interior of K, respectively. Since K is connected and S is smooth, Ω is also connected.

Denote by $k_1(p) \leqslant \ldots \leqslant k_{m-1}(p)$ the principal curvatures of S at $p \in S$ with respect to the normal vector $\nu(p)$ pointing out of K. By Proposition 4.2.6, K is two-convex if and only if $k_2(p) \geqslant 0$ for any $p \in S$.

"Only if" part. Assume K is not two-convex. Then by Proposition 4.2.6, there is a triangle $[xyz]$ in K which is null-homotopic in K, but the solid triangle $\Delta = \operatorname{Conv}\{x, y, z\}$ does not lie in K completely. Evidently the triangle $[xyz]$ is not thin in K. Hence K is not CAT(0).

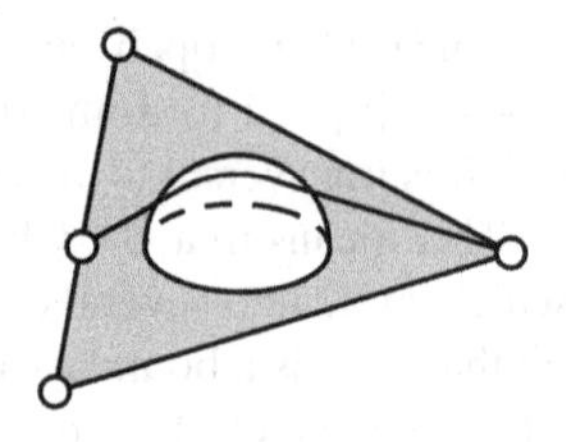

"If" part. Since K is simply connected, by the globalization theorem (3.3.1) it suffices to show that any point $p \in K$ admits a CAT(0) neighborhood.

If $p \in \operatorname{Int} K$, then it admits a neighborhood isometric to a CAT(0) subset of $\mathbb{E}^m$. Fix $p \in S$. Assume that $k_2(p) > 0$. Fix a sufficiently small $\varepsilon > 0$ and set $K' = K \cap \overline{B}[p, \varepsilon]$. Let us show that

❶ K' is CAT(0).

Consider the coordinate system with the origin at p and the principal directions and $\nu(p)$ as the coordinate directions. For small $\varepsilon > 0$, the set K' can be described as a subgraph

$$K' = \left\{ (x_1, \dots, x_m) \in \overline{B}[p, \varepsilon] : x_m \leqslant f(x_1, \dots, x_{m-1} \right\}.$$

Fix $s \in [-\varepsilon, \varepsilon]$. Since ε is small and $k_2(p) > 0$, the restriction $f|_{x_1=s}$ is concave in the $(m-2)$-dimensional cube defined by the inequalities $|x_i| < 2 \cdot \varepsilon$ for $2 \leqslant i \leqslant m-1$.

Fix a negative real value $\lambda < k_1(p)$. Given $s \in (-\varepsilon, \varepsilon)$, consider the set

$$V_s = \left\{ (x_1, \dots, x_m) \in K' : x_m \leqslant f(x_1, \dots, x_{m-1}) + \lambda \cdot (x_1 - s)^2 \right\}.$$

Note that the function

$$(x_1, \dots, x_{m-1}) \mapsto f(x_1, \dots, x_{m-1}) + \lambda \cdot (x_1 - s)^2$$

is concave near the origin. Since ε is small, we can assume that the V_s are convex subsets of $\mathbb{E}^m$.

Further note that

$$K' = \bigcup_{s \in [-\varepsilon, \varepsilon]} V_s.$$

Also, the same argument as in Question 4.1.1 shows that

❷ *If $a < b < c$, then $V_b \supset V_a \cap V_c$.*

Given an array of values $s^1 < \dots < s^k$ in $[-\varepsilon, \varepsilon]$, set $V^i = V_{s^i}$ and consider the unions

$$W^i = V^1 \cup \dots \cup V^i$$

equipped with the induced length metric.

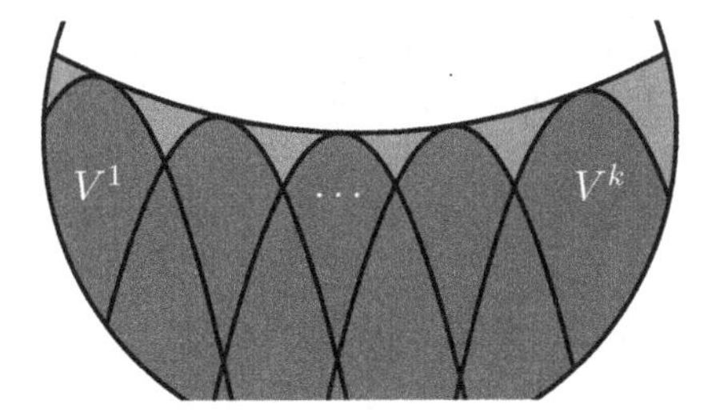

Note that the array (s^n) can be chosen in such a way that W^k is arbitrarily close to K' in the sense of Hausdorff.

By Proposition 2.1.1, in order to prove ❶, it is sufficient to show the following:

❸ *All* W^i *are* CAT(0).

This claim is proved by induction. Base: $W^1 = V^1$ is CAT(0) as a convex subset in $\mathbb{E}^m$.
Step: Assume that W^i is CAT(0). According to ❷,

$$V^{i+1} \cap W^i = V^{i+1} \cap V^i.$$

Moreover, this is a convex set in $\mathbb{E}^m$, and therefore it is a convex set in W^i and in V^{i+1}. By the Reshetnyak gluing theorem, W^{i+1} is CAT(0). Hence the claim follows. △

Note that we have proved the following:

❹ K' *is* CAT(0) *if* K *is* strongly two-convex, *that is,* $k_2(p) > 0$ *at any point* $p \in S$.

It remains to show that p admits a CAT(0) neighborhood in the case $k_2(p) = 0$.

Choose a coordinate system $(x_1, \ldots, x_m)$ as above, so that the $(x_1, \ldots, x_{m-1})$-coordinate hyperplane is the tangent subspace to S at p.

Fix $\varepsilon > 0$ so that a neighborhood of p in S is the graph

$$x_m = f(x_1, \ldots, x_{m-1})$$

of a function f defined on the open ball B of radius ε centered at the origin in the $(x_1, \ldots, x_{m-1})$-hyperplane. Fix a smooth positive strongly convex function $\varphi\colon B \to \mathbb{R}_+$ such that $\varphi(x) \to \infty$ as x approaches the boundary of B. Note that for $\delta > 0$, the subgraph K_δ defined by the inequality

$$x_m \leqslant f(x_1, \ldots, x_{m-1}) - \delta \cdot \varphi(x_1, \ldots, x_{m-1})$$

is strongly two-convex. By ❹, K_δ is CAT(0).

Finally as $\delta \to 0$, the closed ε-neighborhoods of p in K_δ converge to the closed ε-neighborhood of p in K. By Proposition 2.1.1, the ε-neighborhood of p is CAT(0). □

4.4 Open plane sets

In this section we consider inheritance of upper curvature bounds by subsets of the Euclidean plane.

4.4.1. Theorem. *Let Ω be an open simply connected subset of $\mathbb{E}^2$. Equip Ω with its induced length metric and denote its completion by K. Then K is* CAT(0).

The assumption that the set Ω is open is not critical; instead one can assume that the induced length metric takes finite values at all points of Ω. We sketch the proof given by Richard Bishop in [15] and leave the details to be finished as an exercise. A generalization of this result is proved by Alexander Lytchak and Stefan Wenger [46, Proposition 12.1]; this paper also contains a far-reaching application.

Sketch of proof. It is sufficient to show that any triangle in K is thin, as defined in 2.2.1.

Note that K admits a length-preserving map to $\mathbb{E}^2$ that extends the embedding $\Omega \hookrightarrow \mathbb{E}^2$. Therefore each triangle $[xyz]$ in K can be mapped to the plane in a length-preserving way. Since Ω is simply connected, any open region, say Δ, that is surrounded by the image of $[xyz]$ lies completely in Ω.

Note that in each triangle $[xyz]$ in K, the sides $[xy]$, $[yz]$, and $[zx]$ intersect each other along a geodesic starting at a common vertex, possibly a one-point geodesic. In other words, every triangle in K looks like the one in the diagram.

Indeed, assuming the contrary, there will be a *lune* in K bounded by two minimizing geodesics with common ends but no other common points. The image of this lune in the plane must have concave sides, since otherwise one could shorten the sides by pushing them into the interior. Evidently, there is no plane lune with concave sides, a contradiction.

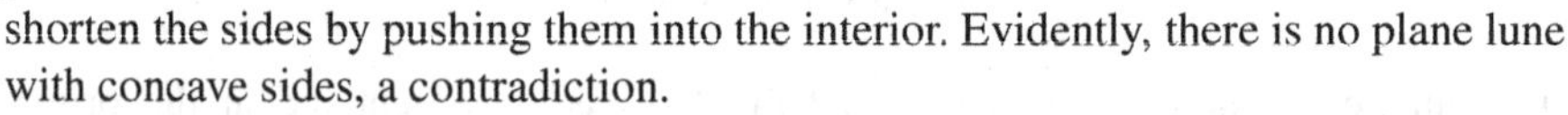

Note that it is sufficient to consider only simple triangles $[xyz]$, that is, triangles whose sides $[xy]$, $[yz]$ and $[zx]$ intersect each other only at the common vertices. If this is not the case, chopping the overlapping part of sides reduces to the injective case (this is formally stated in Exercise 4.4.2).

Again, the open region, say Δ, bounded by the image of $[xyz]$ has concave sides in the plane, since otherwise one could shorten the sides by pushing them into Ω. It remains to solve Exercise 4.4.3. □

4.4.2. Exercise. *Assume that $[pq]$ is a common part of the two sides $[px]$ and $[py]$ of the triangle $[pxy]$. Consider the triangle $[qxy]$ whose sides are formed by arcs of the sides of $[pxy]$. Show that if $[qxy]$ is thin, then so is $[pxy]$.*

4.4.3. Exercise. *Assume S is a closed plane region whose boundary is a triangle T with concave sides in the plane. Equip S with the induced length metric. Show that the triangle T is thin in S.*

Here is a spherical analog of Theorem 4.4.1, which can be proved along the same lines. It will be used in the next section.

4.4.4. Exercise. *Let Θ be an open connected subset of the unit sphere $\mathbb{S}^2$ that does not contain a closed hemisphere. Equip Θ with the induced length metric. Let $\tilde{\Theta}$ be a metric cover of Θ such that any closed curve in $\tilde{\Theta}$ shorter than $2\cdot\pi$ is contractible.*

Show that the completion of $\tilde{\Theta}$ is $\mathrm{CAT}(1)$.

4.5 Shefel's theorem

In this section we will formulate our version of a theorem of Samuel Shefel and prove a couple of its corollaries.

It seems that Shefel was very intrigued by the survival of metric properties under affine transformation.

To describe an instance of such phenomena, note that two-convexity survives under affine transformations of a Euclidean space. Therefore, as a consequence of the smooth two-convexity theorem (4.3.1), the following holds.

4.5.1. Corollary. *Let K be closed connected subset of Euclidean space equipped with the induced length metric. Assume K is* $\mathrm{CAT}(0)$ *and bounded by a smooth hypersurface. Then any affine transformation of K is also* $\mathrm{CAT}(0)$.

By Corollary 4.5.4, an analogous statement holds for sets bounded by Lipschitz surfaces in the three-dimensional Euclidean space. In higher dimensions this is no longer true, see Exercise 4.8.2.

Here is the main theorem of this section.

4.5.2. Two-convexity theorem. *Let Ω be a connected open set in $\mathbb{E}^3$. Equip Ω with the induced length metric and denote by $\tilde{K}$ the completion of the universal metric cover of Ω. Then $\tilde{K}$ is* $\mathrm{CAT}(0)$ *if and only if Ω is two-convex.*

The proof of this statement will be given in the following three sections. First we prove its polyhedral analog, then we prove some properties of two-convex hulls in three-dimensional Euclidean space and only then do we prove the general statement.

The following exercise shows that the analogous statement does not hold in higher dimensions.

4.5.3. Exercise. *Let Π_1, Π_2 be two planes in $\mathbb{E}^4$ intersecting at a single point. Let $\tilde{K}$ be the completion of the universal metric cover of $\mathbb{E}^4\backslash(\Pi_1\cup\Pi_2)$.*

Show that $\tilde{K}$ is $\mathrm{CAT}(0)$ *if and only if $\Pi_1\perp\Pi_2$.*

Before coming to the proof of the two-convexity theorem, let us formulate a few corollaries. The following corollary is a generalization of the smooth two-convexity theorem (4.3.1) for three-dimensional Euclidean space.

4.5.4. Corollary. *Let K be a closed subset in $\mathbb{E}^3$ bounded by a Lipschitz hypersurface. Then K with the induced length metric is* CAT(0) *if and only if the interior of K is two-convex and simply connected.*

Proof. Set $\Omega = \operatorname{Int} K$. Since K is simply connected and bounded by a surface, Ω is also simply connected.

Apply the two-convexity theorem to Ω. Note that the completion of Ω equipped with the induced length metric is isometric to K with the induced length metric. Hence the result. □

Note that the Lipschitz condition is used just once to show that the completion of Ω is isometric to K with the induced length metric. This property holds for a wider class of hypersurfaces; for instance the Alexander horned ball might have CAT(0) induced length metric.

The following corollary is the main statement in Shefel's original paper [62]; it generalizes Alexandrov's theorem about ruled surfaces in [9]. In order to formulate it, we need yet one more definition.

Let U be an open set in $\mathbb{R}^2$. A continuous function $f: U \to \mathbb{R}$ is called *saddle* if for any linear function $\ell: \mathbb{R}^2 \to \mathbb{R}$, the difference $f - \ell$ does not have local maxima or local minima in U. Equivalently, the open subgraph and epigraph of f

$$\{ (x, y, z) \in \mathbb{E}^3 : z < f(x, y),\ (x, y) \in U \},$$
$$\{ (x, y, z) \in \mathbb{E}^3 : z > f(x, y),\ (x, y) \in U \}$$

are two-convex.

4.5.5. Corollary. *Let $f: \mathbb{D} \to \mathbb{R}$ be a Lipschitz function which is saddle in the interior of the closed unit disk $\mathbb{D}$. Then the graph*

$$\Gamma = \{ (x, y, z) \in \mathbb{E}^3 : z = f(x, y) \},$$

equipped with induced length metric is CAT(0).

Proof. Since the function f is Lipschitz, its graph Γ with the induced length metric is bi-Lipschitz equivalent to $\mathbb{D}$ with the Euclidean metric.

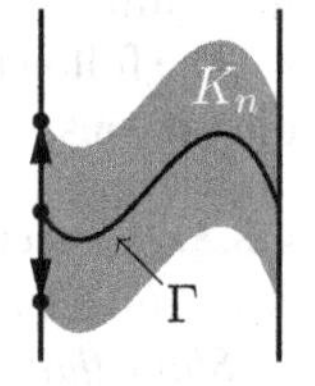

Consider the sequence of sets

$$K_n = \{ (x, y, z) \in \mathbb{E}^3 : z \lessgtr f(x, y) \pm \tfrac{1}{n},\ (x, y) \in \mathbb{D} \}.$$

Note that each K_n is closed and simply connected. By definition K is also two-convex. Moreover the boundary of K_n is a Lipschitz surface.

Equip K_n with the induced length metric. By Corollary 4.5.4, K_n is CAT(0). It remains to note that $K_n \to \Gamma$ in the sense of Gromov–Hausdorff, and apply Proposition 2.1.1. □

4.6 Polyhedral case

Now we are back to the proof of the two-convexity theorem (4.5.2).

Recall that a subset P of $\mathbb{E}^m$ is called a *polytope* if it can be presented as a union of a finite number of simplices. Similarly, a *spherical polytope* is a union of a finite number of simplices in $\mathbb{S}^m$.

Note that any polytope admits a finite triangulation. Therefore any polytope equipped with the induced length metric forms a Euclidean polyhedral space as defined in 3.4.1.

Let P be a polytope and Ω its interior, both considered with the induced length metrics. Typically, the completion K of Ω is isometric to P. However in general we only have a locally distance preserving map $K \to P$; it does not have to be onto and it may not be injective. An example can be guessed from the picture. Nevertheless, is easy to see that K is always a polyhedral space.

4.6.1. Lemma. *The two-convexity theorem (4.5.2) holds if the set Ω is the interior of a polytope.*

The statement might look obvious, but there is a hidden obstacle in the proof. The proof uses the following two exercises.

4.6.2. Exercise. *Show that any closed path of length $< 2 \cdot \pi$ in the units sphere $\mathbb{S}^2$ lies in an open hemisphere.*

4.6.3. Exercise. *Assume Ω is an open subset in $\mathbb{E}^3$ that is not two-convex. Show that there is a plane W such that the complement $W \backslash \Omega$ contains an isolated point and a small circle around this point in W is contractible in Ω.*

Proof of 4.6.1. The "only if" part can be proved in the same way as in the smooth two-convexity theorem (4.3.1) with additional use of Exercise 4.6.3.

"If" part. Assume that Ω is two-convex. Denote by $\tilde{\Omega}$ the universal metric cover of Ω. Let $\tilde{K}$ and K be the corresponding completions of $\tilde{\Omega}$ and Ω.

The main step is to show that $\tilde{K}$ is CAT(0).

Note that K is a polyhedral space and the covering $\tilde{\Omega} \to \Omega$ extends to a covering map $\tilde{K} \to K$ which might be branching at some vertices.[1]

[1]For example, if $K = \{ (x, y, z) \in \mathbb{E}^3 : |z| \leqslant |x| + |y| \leqslant 1 \}$ and p is the origin, then Σ_p, the space of directions at p, is not simply connected and $\tilde{K} \to K$ branches at p.

Fix a point $\tilde{p} \in \tilde{K} \backslash \tilde{\Omega}$; denote by p the image of $\tilde{p}$ in K. Note that $\tilde{K}$ is a ramified cover of K and hence is locally contractible. Thus, any loop in $\tilde{K}$ is homotopic to a loop in $\tilde{\Omega}$ which is simply connected. Therefore $\tilde{K}$ is simply connected too.

Thus, by the globalization theorem (3.3.1), it is sufficient to show that

❶ *a small neighborhood of* $\tilde{p}$ *in* $\tilde{K}$ *is* CAT(0).

Recall that $\Sigma_{\tilde{p}} = \Sigma_{\tilde{p}}\tilde{K}$ denotes the space of directions at $\tilde{p}$. Note that a small neighborhood of $\tilde{p}$ in $\tilde{K}$ is isometric to an open set in the cone over $\Sigma_{\tilde{p}}\tilde{K}$. By Exercise 2.1.2, ❶ follows once we can show that

❷ $\Sigma_{\tilde{p}}$ *is* CAT(1).

By rescaling, we can assume that every face of K which does not contain p lies at distance at least 2 from p. Denote by $\mathbb{S}^2$ the unit sphere centered at p, and set $\Theta = \mathbb{S}^2 \cap \Omega$. Note that $\Sigma_p K$ is isometric to the completion of Θ and $\Sigma_{\tilde{p}}\tilde{K}$ is the completion of the regular metric covering $\tilde{\Theta}$ of Θ induced by the universal metric cover $\tilde{\Omega} \to \Omega$.

By Exercise 4.4.4, it remains to show the following:

❸ *Any closed curve in* $\tilde{\Theta}$ *shorter than* $2 \cdot \pi$ *is contractible.*

Fix a closed curve $\tilde{\gamma}$ of length $< 2 \cdot \pi$ in $\tilde{\Theta}$. Its projection γ in $\Theta \subset \mathbb{S}^2$ has the same length. Therefore, by Exercise 4.6.2, γ lies in an open hemisphere. Then for a plane Π passing close to p, the central projection γ' of γ to Π is defined and lies in Ω. By construction of $\tilde{\Theta}$, the curve γ and therefore γ' are contractible in Ω. From two-convexity of Ω and Proposition 4.2.7, the curve γ' is contractible in $\Pi \cap \Omega$.

It follows that γ is contractible in Θ and therefore $\tilde{\gamma}$ is contractible in $\tilde{\Theta}$. □

4.7 Two-convex hulls

The following proposition describes a construction which produces the two-convex hull $\operatorname{Conv}_2 \Omega$ of an open set $\Omega \subset \mathbb{E}^3$. This construction is very close to the one given by Samuel Shefel in [62].

4.7.1. Proposition. *Let* $\Pi_1, \Pi_2 \ldots$ *be an everywhere dense sequence of planes in* $\mathbb{E}^3$. *Given an open set* Ω, *consider the recursively defined sequence of open sets* $\Omega = \Omega_0 \subset \Omega_1 \subset \ldots$ *such that* Ω_n *is the union of* Ω_{n-1} *and all the bounded components of* $\mathbb{E}^3 \backslash (\Pi_n \cup \Omega_{n-1})$. *Then*

$$\operatorname{Conv}_2 \Omega = \bigcup_n \Omega_n.$$

Proof. Set

❶ $$\Omega' = \bigcup_n \Omega_n.$$

Note that Ω' is a union of open sets, in particular Ω' is open.

Let us show that

❷ $$\mathrm{Conv}_2\,\Omega \supset \Omega'.$$

Suppose we already know that $\mathrm{Conv}_2\,\Omega \supset \Omega_{n-1}$. Fix a bounded component $\mathfrak{C}$ of $\mathbb{E}^3\backslash(\Pi_n \cup \Omega_{n-1})$. It is sufficient to show that $\mathfrak{C} \subset \mathrm{Conv}_2\,\Omega$.

By Proposition 4.2.4, $\mathrm{Conv}_2\,\Omega$ is open. Therefore, if $\mathfrak{C} \not\subset \mathrm{Conv}_2\,\Omega$, then there is a point $p \in \mathfrak{C}\backslash\mathrm{Conv}_2\,\Omega$ lying at maximal distance from Π_n. Denote by W_p the plane containing p which is parallel to Π_n.

Note that p lies in a bounded component of $W_p\backslash\mathrm{Conv}_2\,\Omega$. In particular p can be surrounded by a simple closed curve γ in $W_p \cap \mathrm{Conv}_2\,\Omega$. Since p lies at maximal distance from Π_n, the curve γ is null-homotopic in $\mathrm{Conv}_2\,\Omega$. Therefore $p \in \mathrm{Conv}_2\,\Omega$, a contradiction.

By induction, $\mathrm{Conv}_2\,\Omega \supset \Omega_n$ for each n. Therefore ❶ implies ❷.

It remains to show that Ω' is two-convex. Assume the contrary; that is, there is a plane Π and a simple closed curve $\gamma\colon \mathbb{S}^1 \to \Pi \cap \Omega'$ which is null-homotopic in Ω', but not null-homotopic in $\Pi \cap \Omega'$.

By approximation we can assume that $\Pi = \Pi_n$ for a large n, and that γ lies in Ω_{n-1}. By the same argument as in the proof of Proposition 4.2.7 using the loop theorem, we can assume that there is an *embedding* $\varphi\colon \mathbb{D} \to \Omega'$ such that $\varphi|_{\partial\mathbb{D}} = \gamma$ and $\varphi(D)$ lies entirely in one of the half-spaces bounded by Π. By the n-step of the construction, the entire bounded domain U bounded by Π_n and $\varphi(D)$ is contained in Ω', and hence γ is contractible in $\Pi \cap \Omega'$, a contradiction. □

4.7.2. Key lemma. *The two-convex hull of the interior of a polytope in $\mathbb{E}^3$ is also the interior of a polytope.*

Proof. Fix a polytope P in $\mathbb{E}^3$. Set $\Omega = \mathrm{Int}\,P$. We may assume that Ω is dense in P (if not, redefine P as the closure of Ω). Denote by $F_1, \dots, F_m$ the facets of P. By subdividing F_i if necessary, we may assume that all F_i are convex polygons.

Set $\Omega' = \mathrm{Conv}_2\,\Omega$ and let P' be the closure of Ω'. Further, for each i, set $F_i' = F_i\backslash\Omega'$. In other words, F_i' is the subset of the facet F_i which remains on the boundary of P'.

From the construction of the two-convex hull (4.7.1) we have:

❸ *F_i' is a convex subset of F_i.*

Further, since Ω' is two-convex we obtain the following:

❹ *Each connected component of the complement $F_i \backslash F_i'$ is convex.*

Indeed, assume a connected component A of $F_i \backslash F_i'$ fails to be convex. Then there is a supporting line ℓ to F_i' touching F_i' at a single point in the interior of F_i. Then one could rotate the plane of F_i slightly around ℓ and move it parallelly to cut a "cap" from the complement of Ω. The latter means that Ω is not two-convex, a contradiction. △

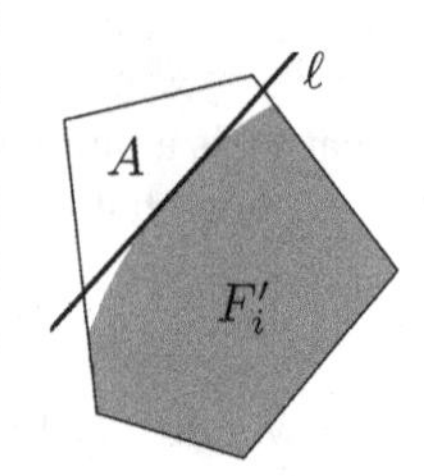

From ❸ and ❹, we conclude

❺ *F_i' is a convex polygon for each i.*

Consider the complement $\mathbb{E}^3 \backslash \Omega$ equipped with the length metric. By construction of the two-convex hull (4.7.1), the complement $L = \mathbb{E}^3 \backslash (\Omega' \cup P)$ is locally convex; that is, any point of L admits a convex neighborhood.

Summarizing: (1) Ω' is a two-convex open set, (2) the boundary $\partial\Omega'$ contains a finite number of polygons F_i' and the remaining part S of the boundary is locally concave. It remains to show that (1) and (2) imply that S and therefore $\partial\Omega'$ are piecewise linear.

4.7.3. Exercise. *Prove the last statement.* □

4.8 Proof of Shefel's theorem

Proof of 4.5.2. The "only if" part can be proved in the same way as in the smooth two-convexity theorem (4.3.1) with the additional use of Exercise 4.6.3.

"If"-part. Suppose Ω is two-convex. We need to show that $\tilde{K}$ is CAT(0).

Fix a quadruple of points $x^1, x^2, x^3, x^4 \in \tilde{\Omega}$. Let us show that CAT(0) comparison holds for this quadruple.

Fix $\varepsilon > 0$. Choose six broken lines in $\tilde{\Omega}$ connecting all pairs of points x^1, x^2, x^3, x^4, where the length of each broken line is at most ε bigger than the distance between its ends in the length metric on $\tilde{\Omega}$. Denote by X the union of these broken lines. Choose a polytope P in Ω such that its interior Int P contains the projections of all six broken lines and disks which contract all the loops created by them (it is sufficient to take 3 disks).

Denote by Ω' the two-convex hull of the interior of P. According to the key lemma (4.7.2), Ω' is the interior of a polytope.

Equip Ω' with the induced length metric. Consider the universal metric cover $\tilde{\Omega}'$ of Ω'. (The covering $\tilde{\Omega}' \to \Omega'$ might be nontrivial—even if Int P is simply connected, its two-convex hull Ω' might not be simply connected.) Denote by $\tilde{K}'$ the completion of $\tilde{\Omega}'$.

By Lemma 4.6.1, $\tilde{K}'$ is CAT(0).

By construction of Int P, the embedding Int $P \hookrightarrow \Omega'$ admits a lift $\iota\colon X \hookrightarrow \tilde{K}'$. By construction, ι almost preserves the distances between the points x^1, x^2, x^3, x^4, namely

$$|\iota(x^i) - \iota(x^j)|_L \gtrless |x^i - x^j|_{\operatorname{Int} P} \pm \varepsilon.$$

Since $\varepsilon > 0$ is arbitrary and CAT(0) comparison holds in $\tilde{K}'$, we get that CAT(0) comparison holds in Ω for x^1, x^2, x^3, x^4.

The statement follows since the quadruple $x^1, x^2, x^3, x^4 \in \tilde{\Omega}$ is arbitrary. □

4.8.1. Exercise. *Assume $K \subset \mathbb{E}^m$ is a closed set bounded by a Lipschitz hypersurface. Equip K with the induced length metric. Show that if K is* CAT(0)*, then K is two-convex.*

The following exercise is analogous to Exercise 4.5.3. It provides a counterexample to the analog of Corollary 4.5.4 in higher dimensions.

4.8.2. Exercise. *Let $K = W \cap W'$, where*

$$W = \left\{ (x, y, z, t) \in \mathbb{E}^4 : z \geqslant -x^2 \right\}$$

and $W' = \iota(W)$ for some motion $\iota\colon \mathbb{E}^4 \to \mathbb{E}^4$.

Show that K is always two-convex and one can choose ι so that K with the induced length metric is not CAT(0).

4.9 Comments

In [1] the first author, David Berg and Richard Bishop gave the exact upper bound on Alexandrov's curvature for the Riemannian manifolds with boundary. This theorem includes the smooth two-convexity theorem (4.3.1) as a partial case. Namely they show the following.

4.9.1. Theorem. *Let M be a Riemannian manifold with boundary ∂M. A direction tangent to the boundary will be called concave if there is a short geodesic in this direction which leaves the boundary and goes into the interior of M. A sectional direction (that is, a 2-plane) tangent to the boundary will be called* concave *if all the directions in it are concave.*

Denote by κ an upper bound of sectional curvatures of M and sectional curvatures of ∂M in the concave sectional directions. Then M is locally CAT(κ).

4.9.2. Corollary. *Let M be a Riemannian manifold with boundary ∂M. Assume that all the sectional curvatures of M and ∂M are bounded above by κ. Then M is locally* CAT(κ).

Under the name *$(n-2)$-convex sets*, two-convex sets in $\mathbb{E}^n$ were introduced by Mikhael Gromov in [38]. In addition to the inheritance of upper curvature bounds by two-convex sets discussed in this chapter, these sets appear as the maximal open sets with vanishing curvature in Riemannian manifolds with nonnegative or nonpositive sectional curvature. This observation was made first by Sergei Buyalo for nonpositive curvature [25, Lemma 5.8], and extended by Dmitri Panov and the third author in [51].

Two-convex sets could be defined using homology instead of homotopy, as in the formulation of the Leftschetz theorem in [38, §$\frac{1}{2}$]. Namely, we can say that K is two-convex if the following condition holds: if a one-dimensional cycle z has support in the intersection of K with a plane W and bounds in K, then it bounds in $K \cap W$.

The resulting definition is equivalent to the one used above. But unlike our definition it can be generalized to define k-convex sets in $\mathbb{E}^m$ for $k > 2$. With this homological definition one can also avoid the use of the loop theorem, whose proof is quite involved. Nevertheless, we chose the definition using homotopies since it is easier to visualize.

Both definitions work well for open sets; for general sets one should be able to give a similar definition using an appropriate homotopy/homology theory.

Let D be an embedded closed disk in $\mathbb{E}^3$. We say that D is *saddle* if each connected bounded component which any plane cuts from D contains a point on the boundary ∂D. If D is locally described by a Lipschitz embedding, then this condition is equivalent to saying that D is two-convex.

4.9.3. Shefel's conjecture. *Any saddle surface in $\mathbb{E}^3$ equipped with the length-metric is locally* CAT(0).

The conjecture is open even for the surfaces described by a bi-Lipschitz embedding of a disk. From another result of Šefel' proved in [61], it follows that a saddle surface satisfies the isoperimetric inequality $a \leqslant C \cdot \ell^2$ where a is the area of a disk bounded by a curve of length ℓ and $C = \frac{1}{3 \cdot \pi}$. By a result of Alexander Lytchak and Stefan Wenger [46], Shefel's conjecture is equivalent to the isoperimetric inequality with the optimal constant $C = \frac{1}{4 \cdot \pi}$.

From Corollary 4.5.4, it follows that if there is a counterexample to Shefel's conjecture, then an arbitrary neighborhood of some point on the surface cannot be presented as a graph in any coordinate system.

Also note that a counterexample could not admit an approximation by smooth saddle surfaces. Thus far, there are no known examples of saddle surfaces which do not admit such approximation.

Note that from the proof of Corollary 4.5.4, it follows that if a two-convex set S can be presented as an intersection of a nested sequence of simply connected open two-convex sets, then S equipped with the length metric is locally CAT(0).

So far there is no known example of a saddle surface which cannot be presented this way. If such a surface S existed, then it would have a point p in its interior and an open set Ω such that

$$\Omega \subset \mathrm{Conv}_2[S \cup \mathrm{B}(p, \varepsilon)].$$

for any $\varepsilon > 0$.

Let us call a subset $K \subset \mathbb{E}^m$ *strongly two-convex* if any null-homotopic circle $\gamma\colon \mathbb{S}^1 \to K$ is also null-homotopic in $K \cap \operatorname{Conv}[\gamma(\mathbb{S}^1)]$.

4.9.4. Question. *Is it true that any closed strongly two-convex set with dense interior bounded by a Lipschitz hypersurface in $\mathbb{E}^m$ is* CAT(0)*?*

More on the subject can be found in the papers [53, 55, 56] by Stephan Stadler and the third author.

Some other constructions of nonpositively curved Riemannian manifolds admit generalizations to CAT spaces: for warped products it was proved by the first author and Richard Bishop [3–5], the conformal deformation was generalized by Alexander Lytchak and Stephan Stadler [47].

Appendix
Semisolutions

Preface

0.0.1. Exercise. Let $\mathcal{X}$ be a 4-point metric space.

Fix a tetrahedron $\triangle$ in $\mathbb{R}^3$. The vertices of $\triangle$, say x_0, x_1, x_2, x_3, can be identified with the points of $\mathcal{X}$.

Note that there is a unique quadratic form W on $\mathbb{R}^3$ such that

$$W(x_i - x_j) = |x_i - x_j|^2_{\mathcal{X}}$$

for all i and j.

By the triangle inequality, $W(v) \geqslant 0$ for any vector v parallel to one of the faces of $\triangle$.

Note that $\mathcal{X}$ is isometric to a 4-point subset in the Euclidean space if and only if $W(v) \geqslant 0$ for any vector v in $\mathbb{R}^3$.

Therefore, if $\mathcal{X}$ is not of type $\mathcal{E}_4$, then $W(v) < 0$ for some vector v. From above, the vector v must be transversal to each of the 4 faces of $\triangle$. Therefore if we project $\triangle$ along v to a plane transversal to v, we see one of the two pictures on the right.

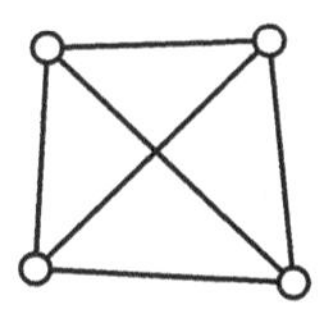
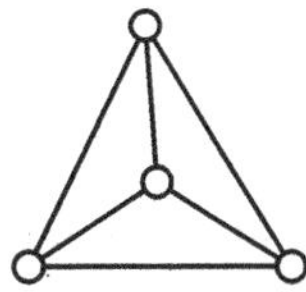

Note that the set of vectors v such that $W(v) < 0$ has two connected components; the opposite vectors v and $-v$ lie in the different components. If one moves v continuously, keeping $W(v) < 0$, then the corresponding projection moves continuously and the projections of the 4 faces cannot degenerate. It follows that the combinatorics of the picture does not depend on the choice of v. Hence $\mathcal{M}_4 \backslash \mathcal{E}_4$ is not connected.

It remains to show that if the combinatorics of the pictures for two spaces is the same, then one can continuously deform one space into the other. This can be easily done by deforming W and applying a permutation of x_0, x_1, x_2, x_3 if necessary. $\square$

The above solution is taken from [54].

S. Alexander et al., *An Invitation to Alexandrov Geometry*,
SpringerBriefs in Mathematics, https://doi.org/10.1007/978-3-030-05312-3

0.0.2. Exercise. The simplest proof we know requires the construction of tangent cones for Alexandrov spaces with nonnegative curvature. (The length metric is defined on page 4.)

Preliminaries

1.1.1. **Exercise.** This exercise is a basic introductory lemma on Gromov–Hausdorff distance (see for example, [20, 7.3.30]). The following proof is not quite standard, and it was suggested to the third author by Travis Morrison.

Given any pair of points $x_0, y_0 \in K$, consider two sequences (x_n) and (y_n) such that $x_{n+1} = f(x_n)$ and $y_{n+1} = f(y_n)$ for each n.

Since K is compact, we can choose an increasing sequence of integers n_i such that both sequences $(x_{n_i})_{i=1}^{\infty}$ and $(y_{n_i})_{i=1}^{\infty}$ converge. In particular, both of these sequences are Cauchy; that is,

$$|x_{n_i} - x_{n_j}|_K, |y_{n_i} - y_{n_j}|_K \to 0 \text{ as } \min\{i, j\} \to \infty.$$

Since f is distance nondecreasing, we get

$$|x_0 - x_{|n_i - n_j|}| \leqslant |x_{n_i} - x_{n_j}|.$$

It follows that there is a sequence $m_i \to \infty$ such that

$$(*) \qquad x_{m_i} \to x_0 \text{ and } y_{m_i} \to y_0 \text{ as } i \to \infty.$$

Set

$$\ell_n = |x_n - y_n|_K.$$

Since f is distance nondecreasing, (ℓ_n) is a nondecreasing sequence.

By $(*)$, $\ell_{m_i} \to \ell_0$ as $m_i \to \infty$. It follows that (ℓ_n) is a constant sequence.

In particular

$$|x_0 - y_0|_K = \ell_0 = \ell_1 = |f(x_0) - f(y_0)|_K$$

for any pair x_0 and y_0. That is, f is distance preserving, in particular, injective.

From $(*)$, we also get that $f(K)$ is everywhere dense. Since K is compact, $f\colon K \to K$ is surjective. Hence the result. □

1.2.1. **Exercise.** A point in $\mathbb{R} \times \operatorname{Cone}\mathcal{U}$ can be described by a triple (x, r, p), where $x \in \mathbb{R}$, $r \in \mathbb{R}_{\geqslant}$ and $p \in \mathcal{U}$. Correspondingly, a point in $\operatorname{Cone}[\operatorname{Susp}\mathcal{U}]$ can be described by a triple (ρ, φ, p), where $\rho \in \mathbb{R}_{\geqslant}$, $\varphi \in [0, \pi]$ and $p \in \mathcal{U}$.

The map $\operatorname{Cone}[\operatorname{Susp}\mathcal{U}] \to \mathbb{R} \times \operatorname{Cone}\mathcal{U}$ defined as

$$(\rho, \varphi, p) \mapsto (\rho \cdot \cos\varphi, \rho \cdot \sin\varphi, p)$$

is the needed isometry. □

1.4.3. **Exercise.** The following example is due to Fedor Nazarov; see [50].

Consider the unit ball (B, ρ_0) in the space c_0 of all sequences converging to zero equipped with the sup-norm.

Consider another metric ρ_1 which is different from ρ_0 by the conformal factor

$$\varphi(\boldsymbol{x}) = 2 + \tfrac{1}{2} \cdot x_1 + \tfrac{1}{4} \cdot x_2 + \tfrac{1}{8} \cdot x_3 + \dots,$$

where $\boldsymbol{x} = (x_1, x_2 \dots) \in B$. That is, if $\boldsymbol{x}(t)$, $t \in [0, \ell]$, is a curve parametrized by ρ_0-length, then its ρ_1-length is

$$\operatorname{length}_{\rho_1} \boldsymbol{x} = \int\limits_0^\ell \varphi \circ \boldsymbol{x}.$$

Note that the metric ρ_1 is bi-Lipschitz to ρ_0.

Assume $\boldsymbol{x}(t)$ and $\boldsymbol{x}'(t)$ are two curves parametrized by ρ_0-length that differ only in the m-th coordinate, denoted by $x_m(t)$ and $x'_m(t)$ correspondingly. Note that if $x'_m(t) \leqslant x_m(t)$ for any t and the function $x'_m(t)$ is locally 1-Lipschitz at all t such that $x'_m(t) < x_m(t)$, then

$$\operatorname{length}_{\rho_1} \boldsymbol{x}' \leqslant \operatorname{length}_{\rho_1} \boldsymbol{x}.$$

Moreover this inequality is strict if $x'_m(t) < x_m(t)$ for some t.

Fix a curve $\boldsymbol{x}(t)$, $t \in [0, \ell]$, parametrized by ρ_0-length. We can choose m large, so that $x_m(t)$ is sufficiently close to 0 for any t. In particular, for some values t, we have $y_m(t) < x_m(t)$, where

$$y_m(t) = (1 - \tfrac{t}{\ell}) \cdot x_m(0) + \tfrac{t}{\ell} \cdot x_m(\ell) - \tfrac{1}{100} \cdot \min\{t, \ell - t\}.$$

Consider the curve $\boldsymbol{x}'(t)$ as above with

$$x'_m(t) = \min\{x_m(t), y_m(t)\}.$$

Note that $\boldsymbol{x}'(t)$ and $\boldsymbol{x}(t)$ have the same end points, and by the above

$$\operatorname{length}_{\rho_1} \boldsymbol{x}' < \operatorname{length}_{\rho_1} \boldsymbol{x}.$$

That is, for any curve $\boldsymbol{x}(t)$ in (B, ρ_1), we can find a shorter curve $\boldsymbol{x}'(t)$ with the same end points. In particular, (B, ρ_1) has no geodesics. □

1.4.7. **Exercise.** The following example is taken from [18].

Consider the following subset of $\mathbb{R}^2$ equipped with the induced length metric

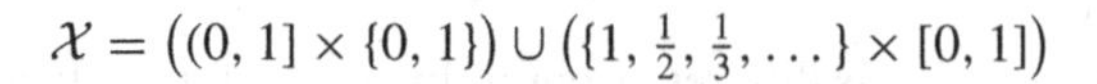

$$\mathcal{X} = \big((0, 1] \times \{0, 1\}\big) \cup \big(\{1, \tfrac{1}{2}, \tfrac{1}{3}, \dots\} \times [0, 1]\big)$$

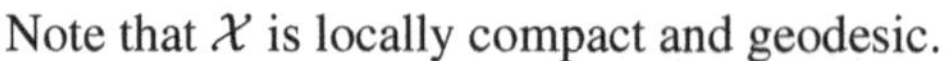

Note that $\mathcal{X}$ is locally compact and geodesic.

Its completion $\bar{\mathcal{X}}$ is isometric to the closure of $\mathcal{X}$ equipped with the induced length metric; $\bar{\mathcal{X}}$ is obtained from $\mathcal{X}$ by adding two points $p = (0, 0)$ and $q = (0, 1)$.

The point p admits no compact neighborhood in $\bar{\mathcal{X}}$, and there is no geodesic connecting p to q in $\bar{\mathcal{X}}$. □

1.6.3. **Exercise.** Assume the contrary; that is

$$\measuredangle[p\,{}^x_z] + \measuredangle[p\,{}^y_z] < \pi.$$

By the triangle inequality for angles (1.6.2) we have

$$\measuredangle[p\,{}^x_y] < \pi.$$

The latter contradicts the triangle inequality for the triangle $[\bar{x} p \bar{y}]$, where the points $\bar{x} \in]\,px]$ and $\bar{y} \in]\,py]$ are sufficiently close to p. □

1.8.2. **Exercise.** By definition of Hausdorff convergence

$$p \in A_\infty \quad \iff \quad \mathrm{dist}_{A_n}(p) \to 0 \quad \text{as} \quad n \to \infty.$$

The latter is equivalent to the existence of a sequence $p_n \in A_n$ such that $|p_n - p| \to 0$ as $n \to \infty$; or equivalently $p_n \to p$. Hence the first statement follows.

The converse is false. For example, consider the alternating sequence of two distinct closed sets $A, B, A, B, \dots$; note that it is not a converging sequence in the sense of Hausdorff. On the other hand, the set of all limit points is well defined—it is the intersection $A \cap B$. □

Remark. The set $\underline{A}_\infty$ of all limits of sequences $p_n \in A_n$ is called the *lower closed limit*, and the set $\bar{A}_\infty$ of all partial limits of such sequences is called the *upper closed limit*. Clearly $\underline{A}_\infty \subset \bar{A}_\infty$. If $\underline{A}_\infty = \bar{A}_\infty$, then it is called the *closed limit* of A_n.[1]

For the class of closed subsets of proper metric spaces, closed limits coincide with limits in the sense of Hausdorff as we defined them.

1.9.4. **Exercise.** To prove part (a), fix a countable dense set of points $\mathfrak{S} \subset \mathcal{X}_\infty$. For each point $x \in \mathfrak{S}$, choose a sequence of points $x_n \in \mathcal{X}_n$ such that $x_n \xrightarrow{\rho} x$.

[1] All these convergences were introduced by Felix Hausdorff in [42].

Applying the diagonal procedure, we can pass to a subsequence of $\mathcal{X}_n$ such that each of the constructed sequences ρ'-converge; that is, $x_n \xrightarrow{\rho'} x'$ for some $x' \in \mathcal{X}'_\infty$.

In this way we get a map $\mathfrak{S} \to \mathcal{X}'_\infty$ defined as $x \mapsto x'$. Note that this map preserves distances and therefore can be extended to a distance-preserving map $\mathcal{X}_\infty \to \mathcal{X}'_\infty$. Likewise we construct a distance preserving map $\mathcal{X}'_\infty \to \mathcal{X}_\infty$.

It remains to apply Exercise 1.1.1.

The proof of part (b) is nearly identical, but one has to apply Exercise 1.1.1 to closed balls centered at the limits of x_n in $\mathcal{X}_\infty$ and $\mathcal{X}'_\infty$. □

Gluing theorem and billiards

2.1.2. **Exercise.** Given a point $x \in \operatorname{Cone}\mathcal{V}$, denote by x' its projection to $\mathcal{V}$ and by $|x|$ the distance from x to the tip of the cone; if x is the tip, then $|x| = 0$ and we can take any point of $\mathcal{U}$ as x'.

Let p, q, x, y be a quadruple in $\operatorname{Cone}\mathcal{V}$. Assume that the spherical model triangles $[\tilde{p}'\tilde{x}'\tilde{y}'] = \tilde{\triangle}(p'x'y')_{\mathbb{S}^2}$ and $[\tilde{q}'\tilde{x}'\tilde{y}'] = \tilde{\triangle}(q'x'y')_{\mathbb{S}^2}$ are defined. Consider the following points in $\mathbb{E}^3 = \operatorname{Cone}\mathbb{S}^2$:

$$\tilde{p} = |p| \cdot \tilde{p}', \qquad \tilde{q} = |q| \cdot \tilde{q}', \qquad \tilde{x} = |x| \cdot \tilde{x}', \qquad \tilde{y} = |y| \cdot \tilde{y}'.$$

Note that $[\tilde{p}\tilde{x}\tilde{y}] \stackrel{iso}{=\!=} \tilde{\triangle}(pxy)_{\mathbb{E}^2}$ and $[\tilde{q}\tilde{x}\tilde{y}] \stackrel{iso}{=\!=} \tilde{\triangle}(qxy)_{\mathbb{E}^2}$. Further note that if $\tilde{z} \in [\tilde{x}\tilde{y}]$, then $\tilde{z}' = \tilde{z}/|\tilde{z}|$ lies on the geodesic $[\tilde{x}'\tilde{y}']$ in $\mathbb{S}^2$. Therefore the CAT(1) comparison for $|p' - q'|$ with $\tilde{z}' \in [\tilde{x}'\tilde{y}']_{\mathbb{S}^2}$ implies the CAT(0) comparison for $|p - q|$ with $\tilde{z} \in [\tilde{x}\tilde{y}]_{\mathbb{E}^3}$.

To show the converse, we need to apply the CAT(0) comparison to a quadruple $s = a \cdot p, q, x, y$ with $a \geqslant 0$ chosen so that the corresponding points $\tilde{s} = a \cdot \tilde{p}, \tilde{q}, \tilde{x}, \tilde{y}$ lie in one plane.

The second statement is proved along the same lines, but we have to use $\mathbb{S}^3 = \operatorname{Susp}\mathbb{S}^2$ instead of $\mathbb{E}^3 = \operatorname{Cone}\mathbb{S}^2$. □

2.1.3. **Exercise.** Fix a quadruple in $\mathcal{U} \times \mathcal{V}$:

$$p = (p_1, p_2), \qquad q = (q_1, q_2), \qquad x = (x_1, x_2), \qquad y = (y_1, y_2).$$

For the quadruple p_1, q_1, x_1, y_1 in $\mathcal{U}$, construct two model triangles $[\tilde{p}_1\tilde{x}_1\tilde{y}_1] = \tilde{\triangle}(p_1x_1y_1)_{\mathbb{E}^2}$ and $[\tilde{q}_1\tilde{x}_1\tilde{y}_1] = \tilde{\triangle}(q_1x_1y_1)_{\mathbb{E}^2}$. Similarly, for the quadruple p_2, q_2, x_2, y_2 in $\mathcal{V}$ construct two model triangles $[\tilde{p}_2\tilde{x}_2\tilde{y}_2]$ and $[\tilde{q}_2\tilde{x}_2\tilde{y}_2]$.

Consider four points in $\mathbb{E}^4 = \mathbb{E}^2 \times \mathbb{E}^2$

$$\tilde{p} = (\tilde{p}_1, \tilde{p}_2), \qquad \tilde{q} = (\tilde{q}_1, \tilde{q}_2), \qquad \tilde{x} = (\tilde{x}_1, \tilde{x}_2), \qquad \tilde{y} = (\tilde{y}_1, \tilde{y}_2).$$

Note that the triangles $[\tilde{p}\tilde{x}\tilde{y}]$ and $[\tilde{q}\tilde{x}\tilde{y}]$ in $\mathbb{E}^4$ are isometric to the model triangles $\tilde{\triangle}(pxy)_{\mathbb{E}^2}$ and $\tilde{\triangle}(qxy)_{\mathbb{E}^2}$.

If $\tilde{z} = (\tilde{z}_1, \tilde{z}_2) \in [\tilde{x}\tilde{y}]$, then $\tilde{z}_1 \in [\tilde{x}_1\tilde{y}_1]$ and $\tilde{z}_2 \in [\tilde{x}_2\tilde{y}_2]$ and

$$|\tilde{z} - \tilde{p}|^2_{\mathbb{E}^4} = |\tilde{z}_1 - \tilde{p}_1|^2_{\mathbb{E}^2} + |\tilde{z}_2 - \tilde{p}_2|^2_{\mathbb{E}^2},$$
$$|\tilde{z} - \tilde{q}|^2_{\mathbb{E}^4} = |\tilde{z}_1 - \tilde{q}_1|^2_{\mathbb{E}^2} + |\tilde{z}_2 - \tilde{q}_2|^2_{\mathbb{E}^2},$$
$$|p - q|^2_{\mathcal{U}\times\mathcal{V}} = |p_1 - q_1|^2_{\mathcal{U}} + |p_2 - q_2|^2_{\mathcal{V}}.$$

Therefore CAT(0) comparison for the quadruples p_1, q_1, x_1, y_1 in $\mathcal{U}$ and p_2, q_2, x_2, y_2 in $\mathcal{V}$ implies CAT(0) comparison for the quadruples p, q, x, y in $\mathcal{U} \times \mathcal{V}$. □

2.1.4. **Exercise.** According to Lemma 1.4.4, it is sufficient to prove the existence of a midpoint between two given points x and y.

For each n choose a $\frac{1}{n}$-midpoint z_n; that is, a point such that

$$|x - z_n|, |y - z_n| \leqslant \tfrac{1}{2} \cdot |x - y| + \tfrac{1}{n}.$$

From the CAT(0) comparison inequality for the quadruple x, y, z_n, z_m we have that $|z_m - z_n| \to 0$ as $n, m \to \infty$; that is, z_n is Cauchy and hence converges since the space is complete.

Let z be the limit of z_n as $n \to \infty$. Clearly z is a midpoint between x and y. □

2.2.8. **Exercise.** Without loss of generality we can assume that $\kappa = 1$. Fix a sufficiently small $0 < \varepsilon < \pi$.

Recall that by Proposition 2.2.7, any local geodesic shorter than π in $\mathcal{U}$ is a geodesic.

Consider a sequence of directions ξ_n at p of geodesics $[pq_n]$. We can assume that the distances $|p - q_n|_{\mathcal{U}}$ are equal to ε for all n; here we use that the geodesics are extendable as local geodesics and minimizing up to length π.

Since $\mathcal{U}$ is proper, the sequence q_n has a partial limit, say q. It remains to note that the direction ξ of $[pq]$ is the limit of directions ξ_n, assuming the latter is defined. □

Note that the unit disk in the plane with attached half-line to each point is a complete CAT(0) length space with extendable geodesics. However, the space of geodesic directions on the boundary of the disk is not complete—there is no geodesic tangent to the boundary of the disk. This provides a counterexample to the statement of the exercise if $\mathcal{U}$ is not assumed to be proper.

2.2.10. **Exercise.** By Alexandrov's lemma (1.5.1), there are nonoverlapping triangles

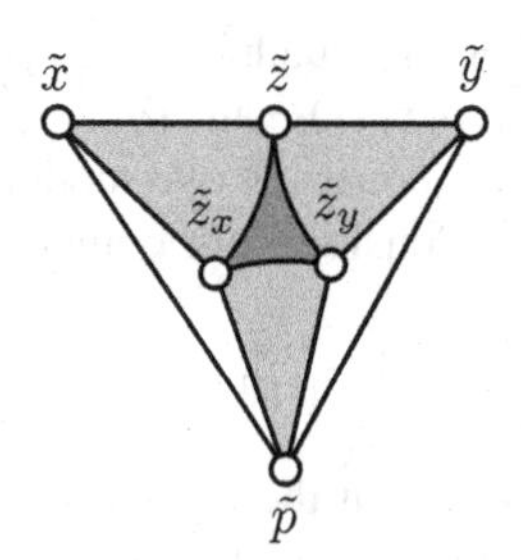

$$[\tilde{p}\tilde{x}\tilde{z}_x] \overset{iso}{=\!=} [\dot{p}\dot{x}\dot{z}]$$

and

$$[\tilde{p}\tilde{y}\tilde{z}_y] \overset{iso}{=\!=} [\dot{p}\dot{y}\dot{z}]$$

inside the triangle $[\tilde{p}\tilde{x}\tilde{y}]$.

Connect the points in each pair $(\tilde{z}, \tilde{z}_x)$, $(\tilde{z}_x, \tilde{z}_y)$, and $(\tilde{z}_y, \tilde{z})$ with arcs of circles centered at $\tilde{y}$, $\tilde{p}$, and $\tilde{x}$ respectively. Define F as follows:

- Map $\mathrm{Conv}[\tilde{p}\tilde{x}\tilde{z}_x]$ isometrically onto $\mathrm{Conv}[\dot{p}\dot{x}\dot{z}]$; similarly map $\mathrm{Conv}[\tilde{p}\tilde{y}\tilde{z}_y]$ onto $\mathrm{Conv}[\dot{p}\dot{y}\dot{z}]$.
- If x is in one of the three circular sectors, say at distance r from its center, set $F(x)$ to be the point on the corresponding segment $[pz]$, $[xz]$ or $[yz]$ whose distance from the left-hand endpoint of the segment is r.
- Finally, if x lies in the remaining curvilinear triangle $\tilde{z}\tilde{z}_x\tilde{z}_y$, set $F(x) = z$.

By construction, F satisfies the conditions. □

2.2.11. **Exercise.** For the CAT(0) case, the statement follows from convexity of distance functions to points in $\mathbb{E}^2$ and thinness of triangles.

For the CAT(1) case, the statement follows from spherical thinness of triangles and convexity of the ball of radius $r < \frac{\pi}{2}$ in $\mathbb{S}^2$.

2.2.12. **Exercise.** Fix a closed, connected, locally convex set K. Note that by Corollary 2.2.5, dist_K is convex in a neighborhood $\Omega \supset K$; that is, dist_K is convex along any geodesic completely contained in Ω.

Since K is locally convex, it is locally path connected. Since K is connected, the latter implies that K is path connected.

Fix two points $x, y \in K$. Let us connect x to y by a path $\alpha\colon [0, 1] \to K$. By Theorem 2.2.3, the geodesic $[x\,\alpha(s)]$ is uniquely defined and depends continuously on s.

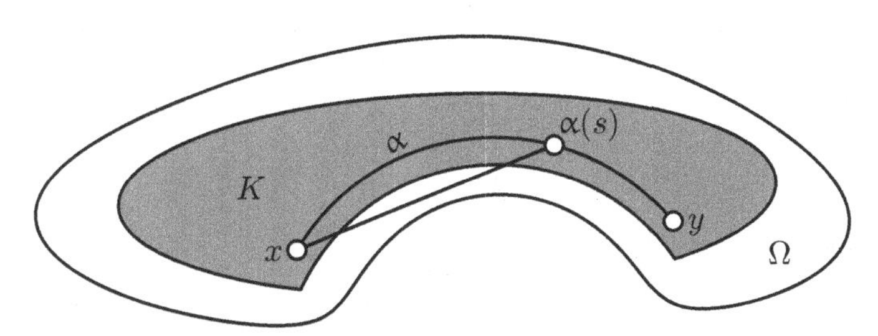

If $[xy] = [x\,\alpha(1)]$ does not completely lie in K, then there is a value $s \in [0, 1]$ such that $[x\,\alpha(s)]$ lies in Ω, but does not completely lie in K. Therefore $f = \mathrm{dist}_K$ is convex along $[x\alpha(s)]$. Note that $f(x) = f(\alpha(s)) = 0$ and $f \geqslant 0$, therefore $f(z) = 0$ for any $z \in [x\,\alpha(s)]$; that is, $[x\,\alpha(s)] \subset K$, a contradiction. □

Comment. The statement generalizes a theorem of Heinrich Tietze, and the proof presented here is nearly identical to the original proof given in [66].

2.2.13. **Exercise.** Since $\mathcal{U}$ is proper, the set $K \cap \overline{\mathrm{B}}[p, R]$ is compact for any $R < \infty$. The existence of at least one point p^* that minimizes the distance from p follows.

Assume p^* is not uniquely defined; that is, two distinct points in K, say x and y, minimize the distance from p. Since K is convex, the midpoint z of $[xy]$ lies in K.

Thinness of triangles implies that

$$|p - z| < |p - x| = |p - y|,$$

a contradiction.

It remains to show that the map $p \mapsto p^*$ is short, that is,

$$|p - q| \geqslant |p^* - q^*|. \qquad ❶$$

for any $p, q \in \mathcal{U}$.

Assume $p \neq p^*$, $q \neq q^*$, $p^* \neq q^*$. In this case, by the first variation inequality (1.6.4),

$$\measuredangle[p^*\,{}^{p}_{q^*}] \geqslant \tfrac{\pi}{2}, \quad \measuredangle[q^*\,{}^{q}_{p^*}] \geqslant \tfrac{\pi}{2},$$

and both angles are defined.

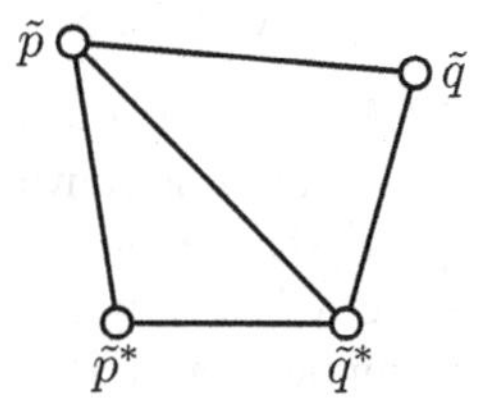

Construct the model triangles $[\tilde p\tilde p^*\tilde q^*]$ and $[\tilde p\tilde q\tilde q^*]$ of $[pp^*q^*]$ and $[pqq^*]$ so that the points $\tilde p^*$ and $\tilde q$ lie on the opposite sides from $[\tilde p\tilde q^*]$.

By comparison,

$$\measuredangle[\tilde p^*\,{}^{\tilde p}_{\tilde q^*}] \geqslant \measuredangle[p^*\,{}^{p}_{q^*}] \geqslant \tfrac{\pi}{2}.$$

Further, applying the triangle inequality for angles (1.6.2) and comparison, we get

$$\begin{aligned}\measuredangle[\tilde q^*\,{}^{\tilde q}_{\tilde p^*}] &= \tilde\measuredangle(q^*\,{}^{q}_{p}) + \tilde\measuredangle(q^*\,{}^{p}_{p^*}) \geqslant \\ &\geqslant \measuredangle[q^*\,{}^{q}_{p}] + \measuredangle[q^*\,{}^{p}_{p^*}] \geqslant \\ &\geqslant \measuredangle[q^*\,{}^{q}_{p^*}] \geqslant \\ &\geqslant \tfrac{\pi}{2},\end{aligned}$$

assuming the left-hand sides are defined. Hence

$$|\tilde p - \tilde q| \geqslant |\tilde p^* - \tilde q^*|.$$

The latter is equivalent to ❶.

In the remaining cases, ❶ holds automatically if (1) $p^* = q^*$ or (2) $p = p^*$ and $q = q^*$. If $p = p^*$, $q \neq q^*$ and $p^* \neq q^*$, then

$$\measuredangle[q^*\,{}^{q}_{p}] \geqslant \tfrac{\pi}{2},$$

and ❶ follows by comparison. □

Comment. This solution leads to the arm lemma in CAT(κ) spaces; see our book [6].

2.4.7. **Exercise.** By approximation, it is sufficient to consider the case when A and B have smooth boundary.

If $[xy] \cap A \cap B \neq \varnothing$, then $z_0 \in [xy]$ and $\dot{A}$, $\dot{B}$ can be chosen to be arbitrary half-spaces containing A and B, respectively.

In the remaining case $[xy] \cap A \cap B = \varnothing$, we have $z_0 \in \partial(A \cap B)$. Consider the solid ellipsoid

$$C = \left\{ z \in \mathbb{E}^m : f(z) \leqslant f(z_0) \right\}.$$

Note that C is compact, convex and has smooth boundary.

Suppose $z_0 \in \partial A \cap \operatorname{Int} B$. Then A and C touch at z_0, and we can set $\dot{A}$ to be the uniquely defined supporting half-space to A at z_0 and $\dot{B}$ to be any half-space containing B. The case $z_0 \in \partial B \cap \operatorname{Int} A$ is treated similarly.

Finally, suppose $z_0 \in \partial A \cap \partial B$. Then the set $\dot{A}$ (respectively, $\dot{B}$) is defined as the unique supporting half-space to A (respectively, B) at z_0 containing A (respectively, B).

Suppose $f(z) < f(z_0)$ for some $z \in \dot{A} \cap \dot{B}$. Since f is concave, $f(\bar{z}) < f(z_0)$ for any $\bar{z} \in [zz_0[$. Since $[zz_0[\ \cap A \cap B \neq \varnothing$, the latter contradicts the fact that z_0 is minimum point of f on $A \cap B$. □

2.5.1. **Exercise.** Fix two open balls $B_1 = \mathrm{B}(0, r_1)$ and $B_2 = \mathrm{B}(0, r_2)$ such that

$$B_1 \subset A^i \subset B_2$$

for each wall A^i.

Note that all the intersections of the walls have ε-wide corners for

$$\varepsilon = 2 \cdot \arcsin \tfrac{r_1}{r_2}.$$

The proof can be guessed from the picture. □

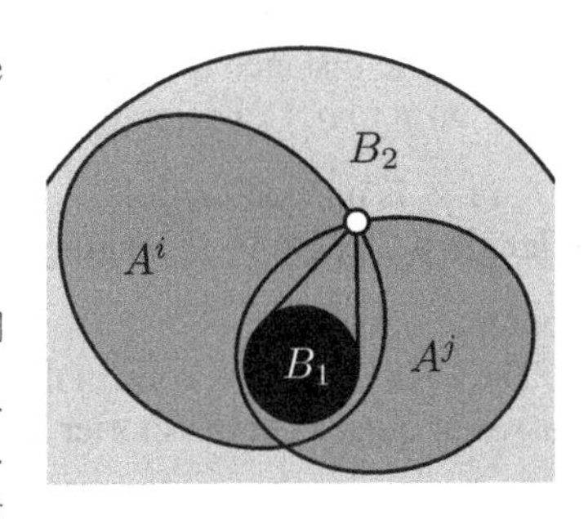

2.5.2. **Exercise.** Note that any centrally symmetric convex closed set in Euclidean space is a product of a compact centrally symmetric convex set and a subspace.

It follows that there is $R < \infty$ such that if X is an intersection of an arbitrary number of walls, then for any point $p \in X$ there is an isometry of X that moves p to a point in the ball $\mathrm{B}(0, R)$.

It remains to repeat the proof of Exercise 2.5.1. □

Globalization and asphericity

3.2.5. **Exercise.** Note that the existence of a null-homotopy is equivalent to the following. There are two one-parameter families of paths α_τ and β_τ, $\tau \in [0, 1]$ such that:

- length α_τ, length $\beta_\tau < \pi$ for any τ.
- $\alpha_\tau(0) = \beta_\tau(0)$ and $\alpha_\tau(1) = \beta_\tau(1)$ for any τ.
- $\alpha_0(t) = \beta_0(t)$ for any t.
- $\alpha_1(t) = \alpha(t)$ and $\beta_1(t) = \beta(t)$ for any t.

By Corollary 3.2.3, the construction in Corollary 3.2.4 produces the same result for α_τ and β_τ. Hence the result. □

3.3.5. **Exercise.** The following proof works for compact locally simply connected metric spaces; by uniqueness of geodesics (2.2.3) this class of spaces includes compact length, locally CAT(κ) spaces.

By the globalization theorem there is a nontrivial homotopy class of closed curves.

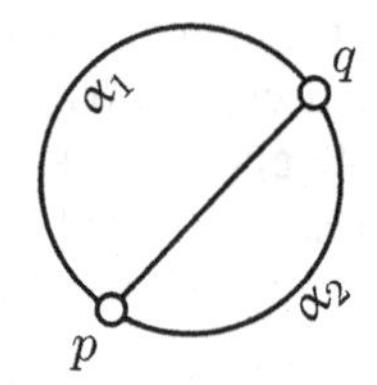

Consider a shortest noncontractible closed curve γ in $\mathcal{X}$; note that such a curve exists.

Indeed, let L be the infimum of lengths of all noncontractible closed curves in $\mathcal{X}$. Compactness and local contractibility of $\mathcal{X}$ imply that any two sufficiently close closed curves in $\mathcal{X}$ are homotopic. Then choosing a sequence of unit speed noncontractible curves whose lengths converge to L, an Arzelá–Ascoli type of argument shows that these curves subconverge to a noncontractible curve of length L.

Assume that γ is not a geodesic circle; that is, there are two points p and q on γ such that the distance $|p - q|$ is shorter then the lengths of the arcs, say α_1 and α_2, of γ from p to q. Consider the products, say γ_1 and γ_2, of $[qp]$ with α_1 and α_2. Then

- γ_1 or γ_2 is noncontractible,
- length γ_1, length $\gamma_2 <$ length γ,

a contradiction.

The CAT(1) case is done in the same way, but we need to consider the homotopy classes of closed curves shorter than $2 \cdot \pi$. □

Note that the statement of the exercise fails if the requirement that $\mathcal{X}$ be compact is replaced by the assumption that it is proper. For example, the surface of revolution of the graph of $y = e^x$ around the x-axis is locally CAT(0) but has no closed geodesics.

3.3.6. **Exercise.** Consider an ε-neighborhood A of the geodesic. Note that A_ε is convex. By the Reshetnyak gluing theorem, the double $\mathcal{W}_\varepsilon$ of $\mathcal{U}$ along A_ε is CAT(0).

Consider the space $\mathcal{W}'_\varepsilon$ obtained by doubly covering $\mathcal{U} \backslash A_\varepsilon$ and gluing back A_ε.

Observe that $\mathcal{W}'_\varepsilon$ is locally isometric to $\mathcal{W}_\varepsilon$. That is, for any point $p' \in \mathcal{W}'_\varepsilon$ there is a point $p \in \mathcal{W}_\varepsilon$ such that the δ-neighborhood of p' is isometric to the δ-neighborhood of p for all small $\delta > 0$.

Further observe that $\mathcal{W}'_\varepsilon$ is simply connected since it admits a deformation retraction onto A_ε, which is contractible. By the globalization theorem, $\mathcal{W}'_\varepsilon$ is CAT(0).

It remains to note that $\tilde{\mathcal{U}}$ can be obtained as a limit of $\mathcal{W}'_\varepsilon$ as $\varepsilon \to 0$ and apply Proposition 2.1.1. □

3.4.3. **Exercise.** Assume $\mathcal{P}$ is not CAT(0). Then by Theorem 3.4.2, a link Σ of some simplex contains a closed geodesic α with length $4 \cdot \ell < 2 \cdot \pi$. Divide α into two equal arcs α_1 and α_2 parametrized by $[-\ell, \ell]$.

Fix a small $\delta > 0$ and consider two curves in Cone Σ written in polar coordinates as

$$\gamma_i(t) = (\alpha_i(\tan \tfrac{t}{\delta}), \sqrt{\delta^2 + t^2}).$$

Observe that both curves γ_1 and γ_2 are geodesics in Cone Σ and have common ends.

Finally observe that a small neighborhood of the tip of Cone Σ admits an isometric embedding into $\mathcal{P}$. Hence the statement follows. □

3.4.4. **Advanced exercise.** Note that it is sufficient to construct a polyhedral space $\mathcal{P}$ homeomorphic to the 3-disk such that (1) $\mathcal{P}$ is locally CAT(0) in its interior and (2) the boundary of $\mathcal{P}$ is locally concave; in particular, each edge on the boundary of $\mathcal{P}$ has angle at least π.

Indeed, once $\mathcal{P}$ is constructed, taking the double of $\mathcal{P}$ along its boundary produces the needed metric on $\mathbb{S}^3$.

The construction of $\mathcal{P}$ goes along the same lines as the construction of a Riemannian metric on the 3-disk with concave boundary and negative sectional curvature. This construction is given by Joel Hass in [41]. □

Note that by the globalization theorem (3.3.1) the obtained metric on $\mathbb{S}^3$ is not locally CAT(0).

This problem originated from a discussion in Oberwolfach between Brian Bowditch, Tadeusz Januszkiewicz, Dmitri Panov, and the third author.

3.5.2. **Exercise.** Checking the flag condition is straightforward once we know the following description of the barycentric subdivision.

Each vertex v of the barycentric subdivision corresponds to a simplex Δ_v of the original triangulation. A set of vertices forms a simplex in the subdivision if it can be ordered, say as $v_1, \dots, v_k$, so that the corresponding simplices form a nested sequence

$$\Delta_{v_1} \subset \dots \subset \Delta_{v_k}.$$ □

3.5.6. **Exercise.** Use induction on the dimension to prove that if in a spherical simplex Δ every edge is at least $\frac{\pi}{2}$, then all dihedral angles of Δ are at least $\frac{\pi}{2}$.

The rest of the proof goes along the same lines as the proof of the flag condition (3.5.5). The only difference is that a geodesic may spend time *at least* π on each visit to Star_v. □

Note that it is not sufficient to assume only that the all dihedral angles of the simplices are at least $\frac{\pi}{2}$. Indeed, the two-dimensional sphere with removed interior of a small rhombus is a spherical polyhedral space glued from four triangles with all the angles at least $\frac{\pi}{2}$. On the other hand the boundary of the rhombus is closed local geodesic in this space. Therefore the space cannot be CAT(1).

3.5.7. **Exercise.** The space $\mathcal{T}_n$ has a natural cone structure with the vertex formed by the completely degenerate tree—all its edges have zero length. Note that the space

Σ over which the cone is taken comes naturally with a triangulation with all-right spherical simplices.

Note that the link of any simplex of this triangulation satisfies the no-triangle condition (3.5.1). Indeed, fix a simplex Δ of the complex; it can be described by combinatorics of a possibly degenerate tree. A triangle in the link of Δ can be described by three ways to resolve a degeneracy by adding one edge of positive length, such that (1) any pair of these resolutions can be done simultaneously, but (2) all three cannot be done simultaneously. Direct inspection shows that this is impossible.

Therefore, by Proposition 3.5.3 our complex is flag. It remains to apply the flag condition (3.5.5) and then Exercise 2.1.2. ☐

3.6.2. **Exercise.** If the complex $\mathcal{S}$ is flag, then its cubical analog $\square_{\mathcal{S}}$ is locally CAT(0) and therefore aspherical.

Assume now that the complex $\mathcal{S}$ is not flag. Extend it to a flag complex $\mathcal{T}$ by gluing a simplex in every clique (that is, a complete subgraph) of its one-skeleton.

Note that the cubical analog $\square_{\mathcal{S}}$ is a proper subcomplex in $\square_{\mathcal{T}}$. Since $\mathcal{T}$ is flag, $\tilde{\square}_{\mathcal{T}}$, the universal cover of $\square_{\mathcal{T}}$, is CAT(0).

Choose a cube Q with minimal dimension in $\tilde{\square}_{\mathcal{T}}$ which is not present in $\tilde{\square}_{\mathcal{S}}$. By Exercise 2.2.12, Q is a convex set in $\tilde{\square}_{\mathcal{T}}$. The closest-point projection $\tilde{\square}_{\mathcal{T}} \to Q$ is a retraction. It follows that the boundary ∂Q is not contractible in $\tilde{\square}_{\mathcal{T}} \backslash \operatorname{Int} Q$. Therefore the spheroid ∂Q is not contractible in $\tilde{\square}_{\mathcal{S}}$. ☐

3.7.3. **Exercise.** The solution goes along the same lines as the proof of Lemma 3.7.2. The only difference is that G is not a subcomplex of the cubical analog. It has to be made from the squares parallel to the squares of the cubical complex which meet the edges of the complex orthogonally at their midpoints. ☐

3.7.5. **Exercise.** In the proof we apply the following lemma. It follows from the disjoint disks property; see [29, 34].

Lemma. *Let $\mathcal{S}$ be a finite simplicial complex which is homeomorphic to an m-dimensional homology manifold for some $m \geqslant 5$. Assume that all vertices of $\mathcal{S}$ have simply connected links. Then $\mathcal{S}$ is a topological manifold.*

Note that it is sufficient to construct a simplicial complex $\mathcal{S}$ such that

- $\mathcal{S}$ is a closed $(m-1)$-dimensional homology manifold;
- $\pi_1(\mathcal{S}\backslash\{v\}) \neq 0$ for some vertex v in $\mathcal{S}$;
- $\mathcal{S} \sim \mathbb{S}^{m-1}$; that is, $\mathcal{S}$ is homotopy equivalent to $\mathbb{S}^{m-1}$.

Indeed, assume such $\mathcal{S}$ is constructed. Then the suspension $\mathcal{R} = \operatorname{Susp} \mathcal{S}$ is an m-dimensional homology manifold with a natural triangulation coming from $\mathcal{S}$. According to the lemma, $\mathcal{R}$ is a topological manifold. According to the generalized Poincaré conjecture, $\mathcal{R} \simeq \mathbb{S}^m$; that is $\mathcal{R}$ is homeomorphic to $\mathbb{S}^m$. Since Cone $\mathcal{S} \simeq \mathcal{R}\backslash\{s\}$ where s denotes a south pole of the suspension and $\mathbb{E}^m \simeq \mathbb{S}^m\backslash\{p\}$ for any point $p \in \mathbb{S}^m$, we get

$$\operatorname{Cone} \mathcal{S} \simeq \mathbb{E}^m.$$

It remains to construct $\mathcal{S}$. Fix an $(m-2)$-dimensional homology sphere Σ with a triangulation such that $\pi_1\Sigma \neq 0$. An example of that type exists for any $m \geqslant 5$; a proof is given in [43].

Remove from Σ the interior of one $(m-2)$-simplex. Denote the resulting complex by Σ'. Since $m \geqslant 5$, we have $\pi_1\Sigma = \pi_1\Sigma'$.

Consider the product $\Sigma' \times [0,1]$. Attach to it the cone over its boundary $\partial(\Sigma' \times [0,1])$. Denote by $\mathcal{S}$ the resulting simplicial complex and by v the tip of the attached cone.

Note that $\mathcal{S}$ is homotopy equivalent to the spherical suspension over Σ, which is a simply connected homology sphere and hence is homotopy equivalent to $\mathbb{S}^{m-1}$. Hence $\mathcal{S} \sim \mathbb{S}^{m-1}$.

The complement $\mathcal{S}\backslash\{v\}$ is homotopy equivalent to Σ'. Therefore

$$\pi_1(\mathcal{S}\backslash\{v\}) = \pi_1\Sigma' = \pi_1\Sigma \neq 0.$$

That is, $\mathcal{S}$ satisfies the conditions above. □

Subsets

4.4.2. **Exercise.** Observe that the triangle $[pqx]$ is degenerate, in particular it is thin. It remains to apply the inheritance lemma (2.2.9).

4.4.3. **Exercise.** By approximation, it is sufficient to consider the case when S has polygonal sides.

The latter case can be done by induction on the number of sides. The base case of triangle is evident.

To prove the induction step, apply Alexandrov's lemma (1.5.1) together with the construction in Exercise 2.2.10. □

4.4.4. **Exercise.** From Exercise 4.6.2, it follows that if a lune in $\mathbb{S}^2$ has perimeter smaller then $2\cdot\pi$, then it contains a closed hemisphere in its interior or lies in an open hemisphere. The same holds for a triangular region with concave sides.

By the assumption, Θ does not contain a closed hemisphere. That is, the first case cannot happen. It remains to apply the argument in the proof of Theorem 4.4.1. □

4.5.3. **Exercise.** The space $\tilde{K}$ is a cone over the branched covering Σ of $\mathbb{S}^3$ infinitely branching along two great circles.

If the planes are not orthogonal, then the minimal distance between the circles is less than $\frac{\pi}{2}$. Assume that this distance is realized by a geodesic $[\xi\zeta]$. The broken line made by four liftings of $[\xi\zeta]$ forms a closed local geodesic in Σ. By Proposition 2.2.7 (or Corollary 3.3.4), Σ is not CAT(1). Therefore by Exercise 2.1.2, K is not CAT(0).

If the planes are orthogonal, then the corresponding great circles in $\mathbb{S}^3$ are subcomplexes of a flag triangulation of $\mathbb{S}^3$ with all-right simplicies. The branching cover is also flag. It remains to apply the flag condition 3.5.5. □

Comments. In [27], Ruth Charney and Michael Davis gave a complete answer to the analogous question for three planes. In particular they show that if a covering space of $\mathbb{E}^4$ branching at three planes intersecting at the origin is CAT(0), then these all are complex planes for some complex structure on $\mathbb{E}^4$.

4.6.2. **Exercise.** Let α be a closed curve in $\mathbb{S}^2$ of length $2\cdot\ell$.

Assume $\ell<\pi$. Let $\check\alpha$ be a subarc of α of length ℓ, with endpoints p and q. Since $|p-q|\leqslant\ell<\pi$, there is a unique geodesic $[pq]$ in $\mathbb{S}^2$. Let z be the midpoint of $[pq]$.

We claim that α lies in the open hemisphere H centered at z.

Assume the contrary; that is, α meets the equator ∂H at a point r. Without loss of generality we may assume that $r\in\check\alpha$.

The arc $\check\alpha$ together with its reflection in z form a closed curve of length $2\cdot\ell$ which meets r and its antipodal point r'. Thus $\ell=\operatorname{length}\check\alpha\geqslant|r-r'|=\pi$, a contradiction. □

Solution via the Crofton formula. Let α be a closed curve in $\mathbb{S}^2$ of length $\leqslant 2\cdot\pi$. We wish to prove that α is contained in a hemisphere in $\mathbb{S}^2$. By approximation it suffices to prove this for smooth curves α of length $<2\cdot\pi$ with transverse self-intersections.

Given $v\in\mathbb{S}^2$, denote by $v^\perp$ the equator in $\mathbb{S}^2$ with the pole at v. Further, $\#X$ will denote the number of points in the set X.

Obviously, if $\#(\alpha\cap v^\perp)=0$, then α is contained in one of the hemispheres determined by $v^\perp$. Note that $\#(\alpha\cap v^\perp)$ is even for almost all v.

Therefore, if α does not lie in a hemisphere, then $\#(\alpha\cap v^\perp)\geqslant 2$ for almost all $v\in\mathbb{S}^2$.

By the Crofton formula we have that

$$\begin{aligned}\operatorname{length}(\alpha)&=\frac{1}{4}\cdot\int\limits_{\mathbb{S}^2}\#(\alpha\cap v^\perp)\cdot \boldsymbol{d}_v\operatorname{area}\geqslant\\&\geqslant 2\cdot\pi.\end{aligned}$$

□

4.6.3. **Exercise.** Since Ω is not two-convex, we can fix a simple closed curve γ that lies in the intersection of a plane W_0 and Ω and is contractible in Ω but not contractible in $\Omega\cap W_0$.

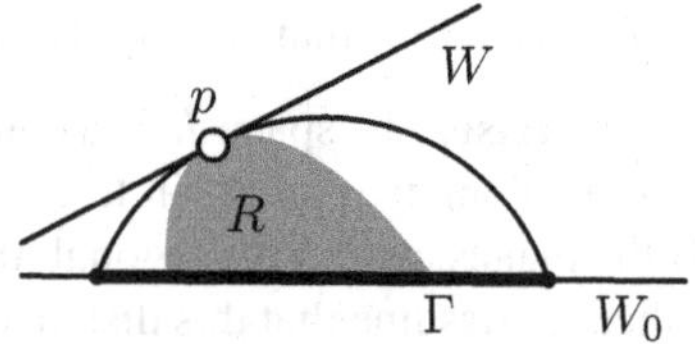

Let $\varphi\colon\mathbb{D}\to\Omega$ be a disk that shrinks γ. Applying the loop theorem (arguing as in the proof of Proposition 4.2.7), we can assume that φ is an embedding and $\varphi(\mathbb{D})$ lies on one side of W_0.

Let Q be the bounded closed domain cut from $\mathbb{E}^3$ by $\varphi(\mathbb{D})$ and W_0. By assumption it contains a point that is not in Ω. Changing W_0, γ, and φ slightly, we can assume that such a point lies in the interior of Q.

Fix a circle Γ in W_0 that surrounds $Q \cap W_0$. Since Q lies in a half-space with boundary W_0, there is a smallest spherical dome with boundary Γ that includes the set $R = Q \backslash \Omega$.

The dome has to touch R at some point p. The plane W tangent to the dome at p has the required property—the point p is an isolated point of the complement $W \backslash \Omega$. Further, by construction a small circle around p in W is contractible in Ω. □

4.7.3. **Exercise.** The proof is simple and visual, but it is hard to write it formally in a nontedious way; for that reason we give only a sketch.

Consider the surface $\bar{S}$ formed by the closure of the remaining part S of the boundary. Note that the boundary ∂S of $\bar{S}$ is a collection of closed polygonal lines.

Assume $\bar{S}$ is not piecewise linear. Show that there is a line segment $[pq]$ in $\mathbb{E}^3$ that is tangent to $\bar{S}$ at some point p and has no common points with $\bar{S}$ except p.

Since $\bar{S}$ is locally concave, there is a local inner supporting plane Π at p that contains the segment $[pq]$.

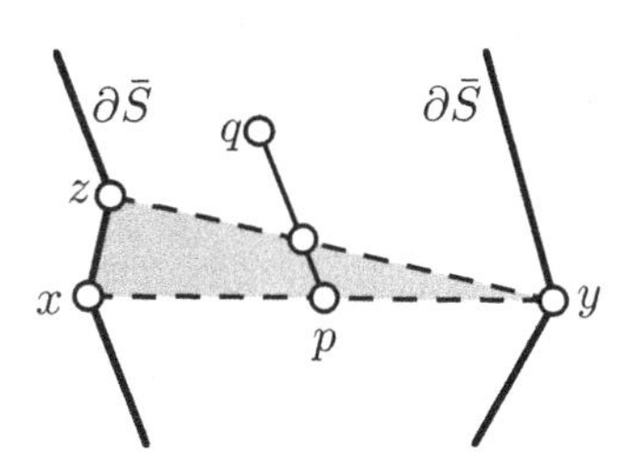

Note that $\Pi \cap \bar{S}$ contains a segment $[xy] \ni p$ with the ends in $\partial\bar{S}$. Denote by Π^+ the half-plane in Π that contains $[pq]$ and has $[xy]$ in its boundary.

Use the fact that $[pq]$ is tangent to S to show that there is a point $z \in \partial\bar{S}$ such that the line segment $[xz]$ or $[yz]$ lies in $\partial\bar{S} \cap \Pi^+$.

From the latter statement and local convexity of $\bar{S}$, it follows that the solid triangle $[xyz]$ lies in $\bar{S}$. In particular, all points on $[pq]$ sufficiently close to p lie in $\bar{S}$, a contradiction. □

4.8.1. **Exercise.** Show that if K is not two-convex, then there is a plane triangle Δ whose sides lie completely in K, but whose interior contains some points from the complement $\mathbb{E}^m \backslash K$.

It remains to note that Δ is not thin in K. □

4.8.2. **Exercise.** Clearly the set W is two-convex. Therefore so is K as the intersection of two two-convex sets.

Consider two 2-dimensional hemispheres H_1 and H_2 in $\mathbb{S}^3$ such that the intersection $H_1 \cap H_2$ is a geodesic $[\xi\zeta]$ orthogonal to the boundary equators of H_1 and H_2 and

$$|\xi - \zeta|_{\mathbb{S}^3} < \tfrac{\pi}{2}.$$

Equip the complement $\mathbb{S}^3 \backslash (H_1 \cup H_2)$ with induced length metric and denote by Σ its completion.

Note that there is a closed geodesic in Σ whose projection to $\mathbb{S}^3$ is formed by a product of four copies of $[\xi\zeta]$. In particular there is a closed geodesic in Σ shorter than $2\cdot\pi$.

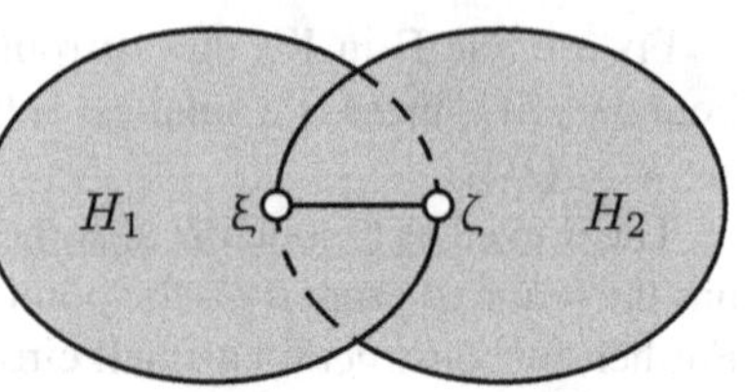

Hence Σ is not CAT(1) and therefore $K' = \operatorname{Cone}\Sigma$ is not CAT(0).

For a suitable choice of the motion ι, we have that $\frac{1}{n}\cdot K \to K'$ as $n\to\infty$ in the sense of Gromov–Hausdorff. Therefore by Proposition 2.1.1, K is not CAT(0). □

References

1. Alexander, S., Berg, D., Bishop, R.: Geometric curvature bounds in Riemannian manifolds with boundary. Trans. Am. Math. Soc. **339**, 703–716 (1993)
2. Alexander, S., Bishop, R.: The Hadamard-Cartan theorem in locally convex spaces. Enseign. Math. **36**, 309–320 (1990)
3. Alexander, S., Bishop, R.: Warped products of Hadamard spaces. Manuscr. Math. **96**(4), 487–505 (1998)
4. Alexander, S., Bishop, R.: Curvature bounds for warped products of metric spaces. GAFA **14**(6), 1143–1181 (2004)
5. Alexander, S., Bishop, R.: Warped products admitting a curvature bound. Adv. Math. **303**, 88–122 (2016)
6. Alexander, S., Kapovitch, V., Petrunin, A.: Alexandrov geometry. http://www.math.psu.edu/petrunin/
7. Александров, А.Д.: *Внутренняя геометрия произвольной выпуклой поверхности* Доклады. Академии наук СССР, **32**(7), 467–470 (1941)
8. Александров, А.Д.: *Одна теорема о треугольниках в метрическом пространстве и некоторые ее прилосжения.* Труды МИАН СССР **38**, 5–23 (1951) (translated into German and combined with more material in [10])
9. Александров, А.Д.: *Линейчатые поверхности в метрических пространствах.* Вестник ЛГУ **2**, 15–44 (1957)
10. Alexandrow, A.D.: Über eine Verallgemeinerung der Riemannschen Geometrie. Schriftenreihe Inst. Math. **1**, 33–84 (1957)
11. Ancel, F., Davis, M., Guilbault, C.: CAT(0) reflection manifolds. Geom. Topol. 441–445
12. Ballmann, W.: Singular spaces of non-positive curvature. Sur les Groupes Hyperboliques d'aprés Mikhael Gromov, Chapter 10. Birkhäuser, Boston (1990) (Progr. Math. **83**)
13. Ballmann, W.: Lectures on Spaces of Nonpositive Curvature. with an appendix by Misha Brin (1995)
14. Billera, L., Holmes, S., Vogtmann, K.: Geometry of the space of phylogenetic trees. Adv. Appl. Math. **27**(4), 733–767 (2001)
15. Bishop, R.: The intrinsic geometry of a Jordan domain. Int. Electron. J. Geom. **1**(2), 33–39 (2008)
16. Blaschke, W.: Kreis und Kugel (1916)
17. Bowditch, B.H.: Notes on locally CAT(1) spaces. In: Geometric Group Theory, pp. 1–48. de Gruyter (1995)
18. Bridson, M., Haefliger, A.: Metric Spaces of Non-positive Curvature (1999)
19. Burago, D.: Hard balls gas and Alexandrov spaces of curvature bounded above. In: Proceedings of the International Congress of Mathematicians, Extra Vol. II, pp. 289–298 (1998) (Vol. II, Doc. Math.)
20. Burago, D., Burago, Yu., Ivanov, S.: A course in metric geometry (2001)

S. Alexander et al., *An Invitation to Alexandrov Geometry*,
SpringerBriefs in Mathematics, https://doi.org/10.1007/978-3-030-05312-3

21. Burago, D., Ferleger, S., Kononenko, A.: Uniform estimates on the number of collisions in semi-dispersing billiards. Ann. Math. **147**(2), 695–708 (1998)
22. Burago, D., Ferleger, S., Kononenko, A.: Topological entropy of semi-dispersing billiards. Ergod. Theory Dyn. Syst. **18**(04), 791–805 (1998)
23. Burago, D., Grigoriev, D., Slissenko, A.: Approximating shortest path for the skew lines problem in time doubly logarithmic in 1/epsilon. Theoret. Comput. Sci. **315**(2–3), 371–404 (2004)
24. Busemann, H.: Spaces with non-positive curvature. Acta Math. **80**, 259–310 (1948)
25. Buyalo, S.V.: Volume and fundamental group of a manifold of nonpositive curvature. Math. USSR-Sbornik **50**(1), 137–150 (1985)
26. Cartan, É.: Leçons sur la Géométrie des Espaces de Riemann (1928)
27. Charney, R., Davis, M.: Singular metrics of nonpositive curvature on branched covers of riemannian manifolds. Am. J. Math. **115**(5), 929–1009 (1993)
28. Charney, R., Davis, M.: Strict hyperbolization. Topology **34**(2), 329–350 (1995)
29. Daverman, R.J.: Decompositions of manifolds (1986)
30. Davis, M.: Groups: generated by reflections and aspherical manifolds not covered by Euclidean space. Ann. Math. **117**(2), 293–324 (1983)
31. Davis, M.: Exotic aspherical manifolds. Topology of high-dimensional manifolds, No. 1, 2, pp. 371–404. Trieste (2001) (ICTP Lect. Notes, 9, 2002)
32. Davis, M., Januszkiewicz, T.: Hyperbolization of polyhedra. J. Differ. Geom. **34**(2), 347–388 (1991)
33. Davis, M., Januszkiewicz, T., Lafont, J.-F.: 4-dimensional locally CAT(0)-manifolds with no Riemannian smoothings. Duke Math. J. **161**(1), 1–28 (2012)
34. Edwards, R.: The topology of manifolds and cell-like maps. In: Proceedings of the International Congress of Mathematicians (Helsinki, 1978), pp. 111–127 (1980)
35. Frolík, Z.: Concerning topological convergence of sets. Czechoskov. Math. J. **10**, 168–180 (1960)
36. Gromov, M.: Hyperbolic groups. Essays in group theory. Math. Sci. Res. Inst. Publ. **8**, 75–264 (1987)
37. Гальперин, Г. А. *О системах локально взаимодействующих и отталкивающихся частиц движущихся в пространстве* Тр. ММО **43**, 142–196 (1981)
38. Gromov, M.: Sign and geometric meaning of curvature (1994)
39. Hadamard, J.: Sur la forme des lignes géodésiques á l'infini et sur les géodésiques des surfaces réglées du second ordre. Bull. Soc. Math. Fr. **26**, 195–216 (1898)
40. Hatcher, A.: Notes on Basic 3-Manifold Topology. A draft is available at http://www.math.cornell.edu/~hatcher/
41. Hass, J.: Bounded 3-manifolds admit negatively curved metrics with concave boundary. J. Differ. Geom. **40**(3), 449–459 (1994)
42. Hausdorff, F.: Grundzüge der Mengenlehre (1914)
43. Kervaire, M.: Smooth homology spheres and their fundamental groups. Trans. Am. Math. Soc. **144**, 67–72 (1969)
44. Menger, K.: The formative years of Abraham Wald and his work in geometry. Ann. Math. Stat. **23**, 14–20 (1952)
45. Lebedeva, N., Petrunin, A.: Local characterization of polyhedral spaces. Geom. Dedicata **179**, 161–168 (2015)
46. Lytchak, A., Wenger, S.: Isoperimetric characterization of upper curvature bounds. arXiv:1611.05261 [math.DG]
47. Lytchak, A., Stadler, S.: Conformal deformations of CAT(0) spaces. Math, Ann (2018)
48. Mazur, B.: A note on some contractible 4-manifolds. Ann. Math. **73**(2), 221–228 (1961)
49. von Mangoldt, H.: Ueber diejenigen Punkte auf positiv gekrümmten Flächen, welche die Eigenschaft haben, dass die von ihnen ausgehenden geodätischen Linien nie aufhören, kürzeste Linien zu sein. J. Reine Angew. Math. **91**, 23–53 (1881)
50. Nazarov, F.: Intrinsic metric with no geodesics. MathOverflow. http://mathoverflow.net/q/15720
51. Panov, D., Petrunin, A.: Sweeping out sectional curvature. Geom. Topol. **18**(2), 617–631 (2014)

52. Panov, D., Petrunin, A.: Ramification conjecture and Hirzebruch's property of line arrangements. To appear in Compositio Mathematica
53. Petrunin, A.: Metric minimizing surfaces. Electron. Res. Announc. Am. Math. Soc. **5**, 47–54 (1999)
54. Petrunin, A.: A quest for 5-point condition a la Alexandrov. St. Petersb. Math. J. **29**(1), 223–225 (2018)
55. Petrunin, A., Stadler, S.: Monotonicity of saddle maps. arXiv:1707.02367 [math.DG] (to appear in Geom. Dedicata)
56. Petrunin, A., Stadler, S.: Metric minimizing surfaces revisited. arXiv:1707.09635 [math.DG]
57. Reshetnyak, Yu.G.: On the theory of spaces of curvature not greater than K. Inextensible mappings in a space of curvature no greater than K. Sib. Math. J. **9**, 683–689 (1968)
58. Reshetnyak, Yu.G.: Two-dimensional manifolds of bounded curvature. In: Geometry, IV. Encyclopaedia Mathematical Sciences, vol. 70, pp. 3–163, 245–250. Springer, Berlin (1993)
59. Rolfsen, D.: Strongly convex metrics in cells. Bull. Am. Math. Soc. **74**, 171–175 (1968)
60. Rinow, W.: Die innere Geometrie der metrischen Raume (1961)
61. Šefel', S.: On saddle surfaces bounded by a rectifiable curve. Soviet Math. Dokl. **6**, 684–687 (1965)
62. Шефель, С.З.: *О внутренней геометрии седловых поверхностей.* Сибирский математический журнал, **5**, 1382–1396 (1964)
63. Stadler, S.: An obstruction to the smoothability of singular nonpositively curved metrics on 4-manifolds by patterns of incompressible tori. GAFA **25**(5), 1575–1587 (2015)
64. Stone, D.: Geodesics in piecewise linear manifolds. Trans. Am. Math. Soc. **215**, 1–44 (1976)
65. Thurston, P.: CAT(0) 4-manifolds possessing a single tame point are Euclidean. J. Geom. Anal. **6**(3), 475–494 (1996) (1997)
66. Tietze, H.: Über Konvexheit im kleinen und im großen und über gewisse den Punkten einer Menge zugeordnete Dimensionszahlen. Math. Z. **28**(1), 697–707 (1928)
67. Wald, A.: Begründung eiiner Koordinatenlosen Differentialgeometrie der Flächen. Ergebnisse eines mathematischen Kolloquium **6**, 24–46 (1935)
68. Wijsman, R.A.: Convergence of sequences of convex sets, cones and functions. II. Trans. Am. Math. Soc. **123**, 32–45 (1966)

Index

S. Alexander et al., *An Invitation to Alexandrov Geometry*,
SpringerBriefs in Mathematics, https://doi.org/10.1007/978-3-030-05312-3

Printed in the United States
By Bookmasters